TECHNOLOGY AND REALITY

Technology and Reality

by
JAMES K. FEIBLEMAN

1982
MARTINUS NIJHOFF PUBLISHERS
THE HAGUE / BOSTON / LONDON

Distributors:

for the United States and Canada

Kluwer Boston, Inc.
190 Old Derby Street
Hingham, MA 02043
USA

for all other countries

Kluwer Academic Publishers Group
Distribution Center
P.O. Box 322
3300 AH Dordrecht
The Netherlands

Library of Congress Cataloging in Publication Data   CIP

Feibleman, James Kern, 1904-
  Technology and reality.

  Includes index.
  1. Technology--Philosophy. I. Title.
T14.F44   601   81-18863
ISBN 90-247-2519-4   AACR2

To J. M. A. Lenihan

TABLE OF CONTENTS

PART TWO. HUMAN NATURE

X

PREFACE

In the following pages I have endeavored to show the impact on philosophy of technology and science; more specifically, I have tried to make up for the neglect by the classical philosophers of the historic role of technology and also to suggest what positive effects on philosophy the almost daily advances in the physical sciences might have. Above all, I wanted to remind the ontologist of his debt to the artificer: technology with its recent gigantic achievements has introduced a new ingredient into the world, and so is sure to influence our knowledge of what there is.

This book, then, could as well have been called '*Ethnotechnology*: An Explanation of Human Behavior by Means of Material Culture', but the picture is a complex one, and there are many more special problems that need to be prominently featured in the discussion.

Human culture never goes forward on all fronts at the same time. In our era it is unquestionably not only technology but also the sciences which are making the most rapid progress. Philosophy has not been very successful at keeping up with them. As a consequence there is an 'enormous gulf between scientists and philosophers today, a gulf which is as large as it has ever been.' (1) I can see that with science moving so rapidly, its current lessons for philosophy might well be outmoded tomorrow. But perhaps the estimates can be made large enough so that gains in philosophy will not be cancelled with every scientific discovery.

The terms I employ in the title are in need of definition. I mean by technology 'the invention and employment of artifacts', and by artifacts the 'materials altered through

1   R.H. Dalitz, in *Nature*, 281, 501 (1979).

human agency for human uses'. I rely on two definitions of reality, ontological and epistemological. The ontological definition is 'equality of being', and the epistemological 'the immediate object of that which is true'.

New Orleans,

June 1979

PART ONE. NATURE

CHAPTER I

INTRODUCTION: THE BACKGROUND

It is a peculiarity of cultural history that technology and science have been neglected by philosophy in quite different ways. Technology was ignored altogether; while the philosophical aspects of science were substituted for the scientific aspects of philosophy.

In the first section of this chapter I propose to consider the importance of technology for philosophy, and in the second what happened to philosophy as a result of its refusal to benefit from the findings of science.

1. Technology Neglected in Theory

Throughout history, technology has played a crucial part in human culture but there is no record that philosophers took any account of it. They give a lot of attention these days to the *existence* of technology but very little to the *products* of technology, but in no major writings of the classical philosophers has the existence of technology even been mentioned. Nowhere, to my knowledge, at least, has there been a philosophy which took the full measure of human practices by assuming that whatever men did about the making and using of artifacts was filled with meanings which needed to be interpreted in philosophical terms.

Philosophers have commented on human behavior, and inferentially on the works of man, for instance architectural buildings have been discussed in theories of aesthetics; but nowhere has there been any mention of the abstract issues raised by the existence of the very visible activities of engineers. True, the results are often taken into account, but never how they came about. It is a cardinal error of philosophy from the very beginning to have ignored the material constructions which make the life of human culture possible.

If man were wholly confined within a natural environment in which there was nothing artificial, it might not be long before he ceased to be anything resembling man. He is dependent for his sheer humanity upon artifacts.

Thus the statement that 'the history of civilization is the history of technology'(1) is not too strong. 'Progress in civilization depends upon invention, and a rapid rate of invention in turn depends upon the sizeable populations that are only possible under civilization.'(2)

The human story began in the Upper Paleolithic some 40,000 years ago with the hunting of big game, standardized tools, personal ornaments and cave wall painting. The earliest civilizations occurred somewhat later in the fertile basins of four rivers: Mesopotamian about 4500 B.C. in the Tigris and Euphrates valleys, Egyptian in the valley of the Nile, the Indus around Harappa and Mohenjo-Daro, and the Chinese near the Yellow River around Anyang. All of them had urban life, writing, large public buildings, and a state organization.

De Camp has argued that the small bands of stone age hunters could have invented very little, and this may explain the static nature of primitive cultures. As soon as city-states were developed and established, and after them the much larger civilizations, the rate of invention increased exponentially. 'Large interconnected populations meant that inventions took place at a faster rate than before, and these inventions in turn made denser and more widely interconnected populations possible. Hence civilized men tended to draw farther and farther ahead of their primitive fellows.'(3)

De Camp's argument is convincing. He pointed out that in the middle of the twentieth century there were 180 million people in the United States, and the Patent Office was issuing 40,000 patents each year. On this basis an isolated band of 45 Americans would produce one invention per century. With two such bands joined together, however, the product would be more than two inventions per century, it would in fact be four. In other words, they would advance technologically twice as fast. And so the larger the society the greater its rate of inventions.

In support of his argument it may be pointed out that capital cities, in which there was a high density of population, were indeed the greatest sources of invention because they made possible a greater interchange of ideas. If this no longer holds true, it is because the enormous increase in the technology of communications weaves investigators no matter where they happen to be working.

Human culture in general is a material product of technology. How much has human life been conditioned by the discovery of printing, gunpowder and the magnet? The

---

1 T.K. Derry and Trevor I. Williams, *A Short History of Technology* (Oxford 1961, University Press), p. 5.
2 L. Sprague de Camp, *The Ancient Engineers* (Garden City, N.Y. 1963, Doubleday), p. 17.
3 *Ibid.*, p. 18.

refusal to recognize this influence, due perhaps to the extreme emphasis on subjective idealism and the insistence, tacit or otherwise, that it consists in ideas and feelings and nothing else, has vitiated all philosophy of culture, including the work of such intuitive masters as Schopenhauer, Hegel and Nietzsche. To none of these would it ever have occurred to suppose that artifacts had any necessary relation to the advance of civilization.

Nietzsche in particular came closer than the others to thinking about the nature of human culture. His philosophy is chiefly devoted to it, and yet he never discussed the material side, without which it could not exist. There are no material tools in Nietzsche's conception, no artifacts, no engineers, no technology, only psychological attitudes, personal drives, subjective outlooks. In a word, no material culture, only the people who produced, maintained and consumed it.

Nietzsche's philosophy would have been a much more satisfactory affair if only he had remembered that culture is not to be found altogether inside the human individual (though it is there, too). Culture is the epigenetic and cumulative collection of artifacts together with the human individuals who know how to use them. It includes beliefs and thoughts, but also everything made by man, from books to buildings and from knitting needles to capital cities. Indeed the evolution of man has been the inevitable result of surrounding himself with that kind of material environment.

Conventional history has been devoted to politics and religion, chiefly an account of wars of conquest as one monarch succeeded another, and altered at times only by the founding of governments and religions and their ascendency over other social forces. If there were palaces and tombs, they were said to have been erected to honor rulers; if there were obelisks and cathedrals, they were built to celebrate saints; if there was opulence and culture, it was ordered by kings and emperors. But the genius of the men who were responsible for these constructions and in fact for everything, from the street lighting which was first introduced in Antioch to the interior plumbing in Mohenjodaro in the Harappa culture of the Indus valley so many thousands of years ago, went unnoticed by the philosophers.

Our records of ancient ages are severely inaccurate and largely incomplete. Literacy, as one critic has suggested, in millennia long past was confined to mere pockets of civilization across the world.(1) There were periods when civilizations reached unheard of heights, such as at Athens in the fifth century B.C., and there were other periods when it touched bottom, as in Europe in the ninth century A.D. Yet all of these developments were ascribed to the successes and failures of intellectual efforts and never credited to the men whose materials constructions — indeed the entire collection of artifacts from drinking cups to river bridges — made everything else possible. The history of human culture, like the history of philosophy, was written by people who

1 Stuart Fleming, in *Science*, 204, 749 (1979).

thought in sedentary terms. Clearly something important is missing from the account.

Who, it might well be asked, built the cities and their contents? Who erected the buildings, laid down the streets, made conduits for water and even lit the main avenues? Who invented writing and improved writing materials, everything from stylus indentations on clay tablets to ink marks on pieces of papyrus reed? Who made paper first from palm fronds and strips of bamboo, and finally from wood pulp and linen rags? Who, in a word, was responsible for the essential material ingredients of civilization?

If there is any one progressive, consistent movement in human history, it is neither political nor religious nor aesthetic. Until recent centuries it was not even scientific. It is the growth of technology, under the guidance of engineers.(1) Capital cities are as old as Sumeria and Egypt. For thousands of years they and others like them have existed without anyone recognizing that they constituted an artificial environment to which eventually man would have to adapt. And they are of course entirely the products of technology.

2. Technology and Its Method

Let me extend the explanation here. I have already defined technology as the production and use of artifacts, and artifacts as materials altered through human agency for human uses. Now I must add that there are two kinds of artifacts: tools and languages. A tool in this broad sense could be anything made by man, from a spade to a skyscraper, from a typewriter to a paved street. A wild horse is not an artifact but a trained horse is. As for language, it is the communication of meaning by means of artifacts, which in this case consists in speaking or writing, the modulation of sound waves or the scratches on a hard surface, respectively. The meaning itself always eventually refers back to some material situation.

And what is the method of technology? What do mechanics, masons, architects, dentists, metallurgists, carpenters, weavers, and many other practitioners of trades and professions, have in common? Is there, in other words, a uniform method in technology as there is in science? Does a technological method exist, somewhat akin to the scientific method? I believe that it does, and that it can be set forth in the following five steps. The technologist has to

1 *Decide what the assignment is*: to build a bridge, a wall, a house; to make a set of dentures, a motor car or a cloth;
2 *Obtain the proper materials*: by reworking clay, shaping bricks, mixing cement, extracting metals, obtaining fibers, felling trees;(2)

---

1 L. Sprague de Camp, *op. cit.*, p. 26.
2 Henry Hodges, *Artifacts* (London 1964, John Baker).

3 *Make the materials if they are not available.* Concrete was a Roman invention. Today it is possible to design and construct materials having almost any of the required properties. The 'materials chemist' tells us he can provide us with any kind of material if we will let him know just what properties we want built into it;(1)

4 *Design the proper tools or machines to perform the task*; whether a mechanical device, such as an engine, a loom or a lever;

5 *Test in order to determine whether the device will work*: whether the conduit will carry water, the pump raise oil, or the engine exert force.

Speaking generally, from the first man who found it necessary to mix and fire clay before it could be used to make a pot, to the man who much more recently had to design the ceramic material for the nose cone of a rocket that must withstand intense heat, the technologist, be he inventor, engineer or applied scientist, ussually worked on the site where the problem presented itself, though this is no longer always the case.

With the oldest human remains have been found crude tools made of bone and stone. Man developed because of his tools; the more complicated the tools, the larger the brain. It was only a few thousand years ago that as a result writing came into existence and along with it the building of cities. Civilization is urbanization; the life of a civilized man consists for the most part in operating whatever tools suit his chosen line of work, everything in fact from a commercial airplane to a concert violin.

The development of material culture went very fast but never for more than a short run of time and always in sudden leaps. The discovery that theory could accelerate practice was perhaps the most recent of the major developments of the human species. Man was able to move back from practice to theory: from invention to engineering to technology to applied science to pure science.

As the complexities of civilization necessitated specialization, the same tasks appeared over and over, and efficiency was increased when they were presented abstractly. The invention of the scientific method of investigation in the seventeenth century brought about a social revolution of the greatest dimensions. Basic researchers were responsible for the discovery of laws, applied scientists used them.

Technology involves inquiry, which is now conducted in the middle of practice. Both applied scientist and technologist perform experiments; but the former do so guided by hypotheses deduced from theory while the latter employ trial-and-error in skilled approaches derived from ordinary experience.

The two streams run together in the engineer, who is a working technologist. In engineering the solutions learned from science and technology are applied to particular cases. Engineering comes down to the business of resorting to techniques and devices to get one piece of matter or energy to move against another piece, which is how man

1   Gerard Piel et. al. (ed.), *Materials* (San Francisco 1967, W.H. Freeman).

exploits the environment to meet his needs. The construction of buildings is facilitated by blue prints. Thus an added degree of abstraction becomes involved when the repeated task discloses the presence of principles which can be learned. The outcome of this tendency is the construction and operation of schools of engineering.

Since the tools of the technologist have become more generally required, it has been discovered that they can be manufactured much in the same way as the objects made by them. In short, factories where machine tools are designed and built have come into existence, and indeed constitute a whole new industry.

No doubt technology has been greatly accelerated in its progress by the assistance of applied science. The designing of new materials has been helped by the knowledge, learned from physics, of their molecular constitution. But on the other hand it is also true that pure science has been aided by technology. The many new instruments which it has made possible have helped accelerate the pace of scientific discovery, as for instance the use of radio telescopes in astronomy.

Instruments have aided greatly in the discovery of laws as well as of facts, and to produce those instruments the services of technologists have been required. The discovery need not have been directly connected with the instruments, indeed the connections are sometimes indirect. Carnot's work (1) in 1824 on the efficiency of heat engines led Lord Kelvin in 1852 to propound the 'universal tendency in nature to the dissipation of mechanical energy' (2), two years later Helmholz generalized to the ultimately stable state of an all-inclusive thermodynamic equilibrium (3) and Clausius finally framed an acceptable description of entropy (4).

The accomplishments of the engineer, though he works on only one project at a time, tend to accumulate. The result is that by now man has entirely transformed his environment. There is nothing within the landscape of a civilized man that was not made by man. The startling fact is that he lives in a wholly artificial world in which everything has been altered, including the air he breathes and the ground he walks on.

He is also engaged though less obviously in the process of transforming himself. A familiar phrase in the accounts of early man is 'man the tool-maker', but there is more to it than that and the phrase can be turned around with greater accuracy. 'Tools the man-maker' would be a better description, for man developed both his large brain and his immense control over the environment by adapting to tools. What man has made, in short, is in the process of remaking man; he must adapt continually to the artificial environment because he continues to change it.

1 E. Mendoza, *Reflections on the Motive Power of Fire by Sadi Carnot* (New York 1960, Dover).
2 W. Thomson, *Phil. Mag.*, 4, 304 (1852).
3 H. von Helmholz, *On The Interaction of Natural Forces* (1854) in M. Kline (ed.), *Popular Scientific Lectures* (New York 1961, Dover).
4 R. Clausius, *Ann. Physik*, 125, 353 (1865); *Phil. Mag.*, 35, 405 (1868).

I have omitted from this account the part played in the development of technology by applied mathematics. 'Applied' mathematics is of course pure mathematics, applied. The missing note in any overly-strict version of radical empiricism is that the knowledge disclosed to sense experience or to instruments has to be mathematically formulated and interpreted before it can be fully utilized. Mathematics enters into every phase of the procedure, yet mathematics is not empirical; it belongs, as we shall see later, to a domain of its own, a domain disclosed to abstract thought but not reducible to it, one which must be discovered rather than invented, for it has its own ground rules and imposes its own liberties and limitations, which are those set forth in formal logic.

Technologists, like the later experimental scientists, never stopped to investigate the assumptions of their procedure. Their method was to be used in particular ways to solve immediate and pressing problems, and they themselves were too busy reshaping materials for the task at hand to be able at the same time to think about what was being taken for granted.

What the technologists have now, however, as the result of purely intuitive procedures, is an elaborate and complex but highly flexible set of techniques for extending the knowledge and control of the material world. The range is enormous, and includes everything from the nearby problem of how to design a new and more efficient kind of electric stove (the microwave oven) to the more remote problem of how to resolve the light from very distant objects situated many light-years from us (the multimirror telescope).

Technology arrived at this stage of development from a start made many millennia ago when it was first discovered that applied mathematics could contribute to architecture, as for instance by the architects Ictinus and Callicrates who designed the Parthenon; and it is progressing much more rapidly in the present epoch, as for instance when scientists at Bell and Howell made the first flexible contact lenses.

Philosophers and other students of human culture are bound to take into account not only the designing and making of artifacts but also their effects. A child grows into an adult with different muscles reinforced by use if he customarily sits in chairs or on the floor. Who is as comfortable with a twisted left arm as a violinist who has practiced from his earliest years? Consider the specialization evident in the mathematician who can deal easily with the details of abstract algebras but not with those of cost accounting. If we add together all the artifacts of a capital city, we must conclude that its inhabitants have been strongly conditioned by their artificial environment, both as to their sense of reality and their determined behavior.

The uses of technology are greatly accelerated by the unity of nature. A good illustration of this is the connection between space science and medicine.

'Fortunately the requirements of other more prosperous industries are very close to the needs of medicine. A piece of machinery which is to be landed on the moon or to be left inside a nuclear reactor must be small in size, light in weight, utterly reliable in operation (since it can never be reached for repair), must consume very small amounts of power and must work well for a long time in an unfriendly environment. These of course are exactly the requirements for equipment which is to be connected to a patient, internally or externally.' (1)

That the requirements of instruments to be employed in the exploration of space are the same as those required to serve in medical technology certainly does point to a nature which is all of a piece.

3. The Rise of Science From Technology

The origins of science are too well known to all to be repeated here. The ancient Greeks knew about the hypothetico-experimental method (2). That they did not get very far with it was only because they failed to comprehend its cumulative nature. Science remained rudimentary because each Greek scientist started anew. It was not until the seventeenth century in Europe that scientist recognized how each generation could build on the achievements of its predecessors; but once they understood that, the progress they made was astonishing.

Another point which is crucial to my thesis is that the problems the early scientists set out to solve were those raised by the philosophers who had been looking at the sky and at their own thoughts before Aristotle developed his own theories about them. Matter and mind began as philosophical concepts and only later were taken up by scientists. For Descartes their nature was of supreme importance even though only because he thought they contributed to an understanding of God.

The points I have been making are of course already well known. What is not so well known, and recited here only as a reminder, is that science was a development of technology. Beginning with tools and instruments constructed for their usefulness in practical tasks, it was only a short logical distance to instruments invented for what they could discover about the nature of things. Galieo's telescope no doubt came about in that way. Science, in a word, was a further extension of technology, and in later centuries proved to be a useful addition to it.

Science may be defined as the experimental search for invariantive laws in nature. Some of those laws have been employed as an aid to technology by facilitating the solution of many problems arising in practice. Science is the kind of technology that addresses itself not to practical problems but to theoretical ones, on the tacit assump-

1 J.M.A. Lenihan, 'The Lessons of History' in Fifty Years of Clinical Physics 1953-2003' (Walton Conference Centre, Southern General Hospital, Glasgow 1978).
2 For further on this method, see below, Chapter V, section 1.

tion that theories discovered in this way will be of immense aid to practice. The search for law by the hypothetico-experimental method was in its origins little more than a kind of anticipatory technology. The success so achieved meant that philosophy was gradually edged out of the picture. Science has benefitted from its origins in technology but by that same token made philosophy seem useless.

That philosophy can still exercise a necessary function is hardly noticed; for the fact is that it cannot supersede the experimental sciences but it can extend them with broader considerations by making connections between them, something no scientst ever does. When the philosopher puts the findings of the sciences together, he is already outside the scientific purview, but has taken himself nearer to a fresh contribution to philosophy.

4. Philosophy in Decline

From the question of the relations of technology to philosophy, we must turn next to look further at that other question which concerns the relations of science to philosophy. It will be one of my theses that the failure to interpret the problems of philosophy by means of the findings of the physical sciences has led to the decline of philosophy. In order to show this, we will need to take a closer look at certain features in the history of the encounters between science and philosophy.

Philosophy on the whole has been more influenced by mathematics than it has by the experimental physical sciences. The direction was probably indicated by logic, which, under the rubric of symbolic or mathematical logic, has become a branch of mathematics rather than a subdivision of philosophy.

The early attempts to develop a philosophical method akin to the scientific method were somewhat naive. The Cartesian method was modeled after the one Descartes found so useful in geometry; but philosophy has a substantive subject-matter, whereas geometry is a branch of mathematics, and as such akin to logic. While the scientific method employs mathematics, it is not the same as the mathematical method prescribed by logic. Spinoza tried to cast his philosophy in mathematical form, but with indifferent results: what value his system has, and I would say it is immense, stems from his philosophical insights which were many, and not from the way in which he linked them together in the mathematical structure made familiar at the time by Euclidean geometry. Spinoza's attempts to perform the same service for Descartes, like the one Bertrand Russell did for Leibnitz, is of lesser value.

The scientific method of investigation consisted in the discovery that advances in engineering could be accelerated by establishing and practicing a formalized procedure — the discovery that there was a method of discovery. Science has added to sense experience many instruments which function as surrogates of the senses. This is the new note introduced into empiricism by science, which enables us to speak in a special way of scientific empiricism.

At first philosophy was very much influenced by science. Kant thought that what he was doing for the humanities was what Newton had done for nature, that the genuine method of metaphysics is fundamentally the same kind which Newton introduced into science and which was there so fruitful (1). He was of course doubly mistaken: he had used not metaphysics but epistemology, and what he had done had absolutely no scientific standing and could offer no scientific analogy.

Not that the assignment was an easy one. The results of the labors of philosophy was thought to be an insight (Descartes) or a series of more or less detached insights (Nietzsche) or a vast compendium (Hegel) or the outline of a set of preconditions (Kant). It may have amounted to no more than a single principle (William of Occam) or it may have represented an attempt to incorporate all of knowledge (Aristotle). It may have been a work of art (Plato) or a pretty dull though valuable business (Dewey).

When philosphers use the term, 'empirical' or 'empiricism', they did not mean the same thing that the experimental scientists meant. Empiricism for the philosopher means being answerable to sense experience; for the scientist it means being answerable to the *disclosures* of sense experience. A world of difference lies between these two meanings. For the philosophers on their interpretation are able to validate empirical statements by reference to their own feelings, whereas the scientists must do so by pointer readings on instruments.

The European philosophers failed because they did not separate the data disclosed to experience from the act of experience by which it was disclosed, thus condemning all knowledge to a permanently subjective status. The scientists succeeded because they did make the separation. They were assisted in making it because they employed instruments to substitute for the senses and read from them data which could hardly have been obtained in any other way.

While Descartes was trying to free his mind from all preconceptions and to make a fresh start in belief with the *cogito, ergo sum* argument as a beginning, Galileo was inventing the telescope and with it finding evidence for the Copernican system. While Spinoza was trying to work out by means of reason alone the modifications of substance, Boyle was conducting experiments which led to the discovery that if the temperature of a gas is constant, the pressure is proportional to the density. And while Leibniz was trying to invoke God in aid of his principle of pre-established harmony, Newton was occupied in showing by means of the prism that lights of different colors were all contained in white light.

When the revolution in physics at the turn of the twentieth century was still fresh in everyone's mind, the astronomer, A.S. Eddington tried to fill a need in a book entitled *The Nature of The Physical World*, which were the Gifford Lectures he delivered at the

---

1   Kant, *Critique of Practical Reason*, III, Introduction.

University of Edinburgh in the winter of 1927. He was very explicit about it in his Preface. At the beginning he explained, 'This book ... treats of the philosophical outcome of the great changes of scientific thought which have recently come about.' In other words, he thought he would have to do the philosopher's work for them by showing the philosophical aspects of the new scientific developments, because there was no comparable effort being conducted by the philosophers. There had been one such effort several years earlier, Whitehead's Lowell Lectures delivered in Boston in 1925 and entitled *Science and The Modern World*, but evidently Eddington did not know about it.

Much of the great progress in science has come about since Whitehead's day in the second and third quarters of the century. After the revolution in physics prompted by the work of Planck, Einstein and others, thousands of books were published explaining the philosophical aspectsof their discoveries, but very few on the other side. In short, there have been many interpretations of science in philosophical terms but almost none of philosophy in scientific terms, a curious area of neglect.

So far as I can discover the situation remains much the same today. Philosophers are still trying to interpret physics in the light of philosophy, not philosophy in the light of physics.

This situation did not come about by accident. It had already been proposed by the positivists that philosophy should be a commentary on science and nothing more. Positivism itself was first suggested perhaps by the scepticism of Hume. He did not hold a positive philosophy by that name but he certainly did advocate the position. As early as 1741, more than half a century before Comte was born, in that now famous passage he had declared in his *Enquire Concerning Human Understanding*, 'If we take in our hand any volume; of divinity or school metaphysics, for instance, let us ask, Does it contain any abstract reasoning concerning quantity or number? No. Does it contain any experimental reasoning matter of fact and existence. No. Commit it to the flames: for it can contain nothing but sophistry and illusion'.

This is not a statement of scepticism but of positivism. Hume's doubt was not directed at mathematics and experimental science, only at philosophy.

Positivism as a philosophy was introduced by August Comte. The first volume of his *Cours de Philosophie Positive* appeared in 1830, the sixth and last volume twelve years later. His position was clearly stated: the only valid knowledge is scientific knowledge, which is not an explanation of nature, only a description of it, but a description in useable practical form. 'Any proposition which does not admit of being ultimately reduced to a simple enunciation of fact, special or general, can have no real or intelligible sense.' John Stuart Mill soon echoed these sentiments. 'The laws of phenomena are all that we know respecting them. Their essential nature and their ultimate causes, either efficient or final, are unknown and inscrutable to us'.

The position was revived and greatly extended in the 1920s by a group of philosophers who described themselves as logical positivists and who came to be known as the members of the 'Vienna Circle'. They rejected metaphysics *in toto* and sought to produce a unification of the sciences without it. They ruled out all knowledge except scientific knowledge. Their principle of procedure was the verification principle, that 'the meaning of a statement is the method of its verification'.

(They chose an unfortunate example, incidentally, for, they argued, the statement 'There are craters on the other side of the moon' was meaningless because there was not and could not be a method of verification. Before many years had passed, men walked on that side of the moon and were able to observe directly the existence of craters.)

The unproductive nature of philosophy compared with the productive nature of science seemed more and more obvious, but that this need not be the case was not evident. The first impact of science on philosophy, then, was to do away with philosophy altogether. This certainly has been the program suggested by Comte and announced explicitly and formally by the logical positivists. It ended with Wittgenstein, whose followers far from obeying his advice to say nothing that could not be said by natural science (1) have been talking volubly ever since.

Positivism as a professional philosophy soon shrank in importance because it was too narrow in scope. If it is true, as positivism avers, that philosophy has nothing of its own to contribute and can only point with pride to the achievement of the scientists, then it is not needed: there are plenty of others who can perform that service.

In the United States the anti-philosophical effects of positivism were reinforced by the native development of pragmatism. Pragmatism as proposed by Peirce was based on the maxim that what is true will work, *a theory of the application of theory to practice*; but his friend, James, misunderstood him, and so advocated the position that what works is true, *a theory of the derivation of theory from practice*. The former is in accord with the best scientific procedure, but the version advocated by James fitted better with the already-prevailing American philosophy.

Frederick Jackson Turner pointed out (2) that the morality of the ever-advancing western frontier in the eighteenth and nineteenth centuries was one that put workability ahead of principle; rough frontier justice, it was called. Trees had to be cut down and cities built, and the main thing was to get on with the job and not let principles stand in the way; if they did, so much the worse for the principles.

James was only saying abstractly what everyone already followed concretely, and so his view prevailed. He will always be welcome who confirms to the members of a popular audience the truth of what they already believe. Nobody troubled to point out that it matters very much when you date workability. The American businessmen declared

1 Ludwig Wittgenstein, *Tractatus Logico-Philosophicus*, 6.53-7.
2 *The Significance of Sections in American History*, with an Introduction by Max Farrand (New York 1932, H. Holt and Company).

Italian fascism a success because in the 1920s Mussolini had the trains running on time. That was before the fascists were defeated and he was hung by his heels when caught trying to escape from Italy. In any case, pragmatism remained strong in the United States where it reinforced the native positivism, from which the prestige of philosophy has yet to recover.

In rejecting philosophy the general public needed no help from the positivists. The typically enlightened individual had already decided that because philosophers use ordinary words in extraordinary ways, he could make no sense of what they said and wrote; at the same time he felt that he need not bother because he was not missing anything important.

Students continue to flock to philosophy courses in universities, but they seem to think of it as an ameliorative cultural topic, much as they might regard a course in Shakespeare or in the history of art: one not expected to have any serious effects on their practical lives. For practice itself they look to business and engineering schools. These, I hasten to point out, are valid pursuits having a greater intellectual and cultural content than has been credited to them, but they are not sufficient in themselves; philosophies are still needed by societies.

Philosophy is a powerful but perishable growth. We have seen that it can be destroyed by an anti-philosophical philosophy like positivism, but it can be destroyed just as easily by the absolute belief in an official philosophy, such as that of Islam. It can flourish only in an atmosphere favorable to inquiry. Thus while it is barely tolerated in the free world as an outmoded enterprise having little or no relevance, it is excluded altogether from communist and dogmatically religious dictatorships.

If members of the generally enlightened public would only look about them, they would see the extent to which they are involved in philosophies, though in a form appearing not as philosophies at all but as settled and absolute political systems. The effect of the work of Locke and Montesquieu on the makers of the American Declaration of Independence and Constitution, and the effect of the work of Marx, Engels and Lenin on the governments of the Soviet Union and Communist China, certainly ought to be sufficient to demonstrate to all concerned the power of philosophy; and ought to convince them, moreover, that a covert and implicit philosophy is far more effective and potentially dangerous than one that is overt and explicit. The metaphysical assumptions of political systems (often avowed by their founders but neglected by their followers) are no less powerful because they go unrecognized.

This is the new and the modern thing: to receive a philosophical system under cover of endorsing a political system, as though theories of reality were so shadowy and evanescent as to be unworthy of being counted at all. When people think about politics or religion, it is in terms of established states or churches, not in terms of the philosophical ideas upon which they were founded. The same people who shrink from any mention of explicit metaphysics are the ones who feel safe inside a government which can justify embodying a metaphysics so long as it is called something else, Thomism or Marxism, for instance.

Here we can see clearly, then, the general effectiveness of the positive philosophy. It has won supremacy in western civilization, and no one understands the necessity either of questioning its validity or of crusading against it. The philosopher is a man wholly without influence, and he has gained this unenviable position for himself by turning aside from philosophy to concentrate on the technical study of language; not its references, mind you, which would take him outside language to the world, but only its meanings, which enables him to stay inside language, a kind of new scholasticism which leaves the real world to the care of others.

5. Prospects for A Philosophical Revival

In the following pages I hope to suggest how philosophy could be revitalized by a fresh approach. What happens to the classical topics in philosophy when they are confronted with the new data derived from observations and experiments in the sciences? Technology has always been a crucial factor in human life but one not noticed by the philosophers. I want to work at correcting this. My aim has been to see what changes are demanded in the conventional and traditional positions of philosophy by the effects of technology and by recent discoveries in the experimental sciences. Thus one of the problems I have encountered is the result of having to fit fresh information into some very old categories. To a great extent new categories are called for, and a different organization; but not altogether, or the aim of this work would have been frustrated entirely.

To argue successfully that philosophy ought to be restored to its ancient greatness, I must show that it has important work to do. It has always kept in view the core of reality, but this somehow got lost when it went its own way independently of the sciences. It had never before ignored such an important source of knowledge. Now it must employ that knowledge, as it formerly did, by restoring the vision by which men live when they establish and direct societies. For philosophy is a way of looking at the world. It can be tested by asking whether it contains inconsistencies and whether anything important has been left out.

Seldom was the hypothesis of a total explanation as urgently called for as it is in our own day. The inventions of the technologists and the discoveries of the scientists have been pouring out in a bewildering profusion for quite a while now, and the confused results cry out for a suggestion of some kind of order. A system of philosophy brought up to date which could take advantage of the important findings is a pressing need. It must be broader and deeper than those findings if it is to include them and yet not be limited by them; but before such a system can be constructed, the impact of technology and science on philosophy must first be ascertained and evaluated. Not 'What does technology and science mean in philosophical terms?' but 'What does philosophy mean in technological and scientific terms?' This will call for illuminating some of the classical problems in philosophy by the discoveries of empiricism.

It could not be more timely. The impoverishing effect of popular positivism has left a bewildered public reaching out for all sorts of substitutes to fill a lack which it feels keenly. In addition to the vogue for astrology, there is the even greater vogue for oriental religions, where philosophy is strong but manages to remain adequately concealed behind all sorts of mystic subterfuges, occult practices and unfamiliar rituals.

Thus philosophy, thrown out of the front door, has quietly ridden back in through a side entrance, this time disguised as something else. The effect has not been altogether favorable to reason and fact, though its emotional impact is undeniable. All the more justification, then, for returning to the high road of speculation via a scientific interpretation of philosophy which can bring to it larger horizons and a deeper understanding of existence.

What is reality? What kind of morality is justified? What turn should religious faith take? So long as we are human and live in the world, we are sure to want answers to these and other questions, and we want moreover the best ones that the available information can supply. They will not be absolute and final answers − perhaps the day for them is gone forever − but more reliably they may be tentative and interim proposals, and our readiness to abandon them for other and better suggestions can be read as a sign of progress. The reality in the present context is equivalent to the demands put on the imagination by empiricism that it limit itself to working over the disclosures of sense experience.

The tasks of philosophy remain, then; and if there is nobody to see them in these terms, then so much the worse; and if no one ever returns to pick up the pieces and make a new beginning, that will be worse still. But the lack would not alter the need for understanding, merely insure that it would continue. And it would be felt as a lack by the individual who might not be aware in explicit formulations of what he was missing yet would be plagued by the lost feeling of having no landmarks. For it is a prerequisite of the enjoyment of life that if its intensity cannot be continued without interruption, at least it is attainable at times. And it can be more reliable if it has the backing of reason.

What should philosophy set out to accomplish these days? Perhaps it is as true for philosophy as it is for technology and science that there is no single enterprise, only a collection of enterprises having a method and an aim in common. Technology may set out to improve the plumbing or the typewriter, science may investigate the solar system or the interior of the biological cell. Philosophy may devise an acceptable morality or construct a philosophy sufficiently wide to accommodate the sciences.

But this would still not be philosophy in the grand sense. It would still be necessary to pool all our information from whatever source and gather it together into one consistent and comprehensive system unified by a single set of axioms and theorems. Nothing less could satisfy the deep human need to know what it is all about, nothing more modest in its claims and explanations would be likely to satisfy our inheritance.

From this last alternative we may take heart that the great tradition is not dead. The aim of philosophy remains what it was in ancient Greece, namely, somehow to describe how all knowledge fits into a system, though now it would have to be a partially-ordered and permanently open system because knowledge from the experimental physical sciences continues to increase exponentially.

More specifically, the task of philosophy is to seek out the presuppositions in any undertaking and try to determine to what extent they have been treated as axioms: how consistent they are, what theorems have flowed from them, and to what extent all of them have been used. In short, anything that exists can be treated as though it were the elements of a system with all of its logical requirements.

This will be especially true of man-made and man-discovered systems, from experimental results in the physical sciences to speculative results in the humanities. And where are we most likely to find the core of such presuppositions? Perhaps in the unconsciously held beliefs about reality, in the arrangement of institutions based upon the priority assigned by such reality, and in the sets of preferences which make themselves felt as values throughout the society.

In the following chapters I have chosen to examine such broad philosophical topics as matter, mind, universals, quality, man, morality, rhetoric, art, politics, religion and nature, in order to determine what changes, if any, technology and science can make in them. The list is intended to be suggestive rather than exhaustive, and I hope that other investigators will be encouraged not only to correct my findings but also to extend them.

CHAPTER II

MATTER

1. Conceptions of Matter, Old and New

Among the many radical changes that have been suggested to philosophy by science, none could be more important than the one that has a bearing on the theory that matter is the primary reality.

Materialism as a metaphysical theory has been around for a long time and has either been accepted or rejected, but in any case never demonstrated. Indeed no theory of metaphysics could be demonstrated, for from what first principles could it be deduced? Arguments for the validity of metaphysical theories must rest on their explanatory value. Positivists ruled out metaphysics on empirical grounds: it has no sense reference, and therefore is non-sense in the literal meaning of the term. That does not keep it from being assumed by other disciplines, however; in fact a metaphysics was assumed inadvertently when physicists experimented with the constituents of matter in subatomic fields as well as when it examined the gross properties of matter in the vast aggregations observed by astronomers, and especially when it tried to put the two sets of findings together in a unified field theory.

That materialism is a theory of metaphysics has sometimes been forgotten. Perhaps the neglect has come about because it has been assumed mistakenly that there could be only one kind of metaphysics. When the metaphysics of idealism is equated with metaphysics itself, we are witnessing the error of substituting one member of a class for the class. We see the same error when subjectivity as a basis for knowledge theory is allowed to preempt the topic of epistemology. Anyone who has ever counted matter among the primary realities is certainly thinking in metaphysical terms.

Another and very special reason for not including the theory of matter in philosophical speculations may have been that a particular brand of materialism has been held to be the absolute truth in Marxist countries, especially in the Soviet Union and Communist China and their satellites; but I can see no good reason for surrendering materialism as a generic classification to the Marxists. There are many new versions of materialism, but the Marxist version is still substantially the old and outdated one Feuerbach taught. He explained reality as founded on sensibility, which, he insisted, is the sole reliable guide. Only through the senses, he held, is an object given; where there is no sense experience there is no real object. Feuerbach was not so much concerned with what the senses disclose as that it is the senses that do the disclosing.

Feuerbach's version was somewhat removed from the strict reading of materialism that the Greek atomists had begun. It was not matter which obsessed Feuerbach but matter-as-known. In this connection he was a true child of the Germany which produced Kant. The earlier tradition of materialism inherited from the Greek atomists actually had been somewhat better than this. According to Leucippus and Democritus the 'a-toms' were hard round bits of stuff, simple, solid and impenetrable. All were falling, with the heavier falling faster and colliding with the lighter and, when not directly in the line of centers, producing a sideways motion.

There were of course limitations to the explanatory value of both conceptions. For a sudden and rapid increase in the knowledge of matter came when physicists in the twentieth century made enormous strides in the discovery of the structure and properties of matter. This was possible because of the rapid advances in technology; the instruments were available to probe the atom and discover its constituents. The results of the inventions of the various devices available to the experimental physicist, everything from the bubble chamber to the linear cyclotron, were very surprising and could in no way have been anticipated. Speaking generally, technology is responsible for the devices which use energy to move matter or alter it in various ways. As a consequence matter is no longer considered as simple inert stuff which resists analysis but has come to be recognized as a highly dynamic agent capable of sustaining the most complex of activities. It can never again be thought of without energy.

Let us turn now to see just what are some of the philosophical aspects of matter which the physicists have led us to accept.

2. The Properties of Matter

In the effort to determine what the new scientific conception of matter suggests in philosophical terms, it may be best to agree about a starting-point. Accordingly, I propose a definition of matter to be employed in the following pages. Since matter is a kind of substance, I will have to offer first a definition of substance.

Accordingly, I define substance as the irrational ground of individual reaction. It has two subdivisions: matter and energy. Matter is static substance; energy, dynamic substance. They are interconvertible according to the well-known Einstein formula, $E=mc^2$. The first and most important property of matter, then, is that it consists in locked-up energy.

It will be best to look first at matter and then at energy, remembering always that we are talking about two modes of the same substance.

When we draw on the physics of the last few decades, we will find that we are justified in concluding that matter has at least six broad properties; it is (a) complex, (b) volatile, (c) divisible, (d) distributed, (e) rare and (f) uniform. A few words of description about each of these will be in order.

(a) The model of the atom offers considerable evidence for the contention that matter is exceedingly complex. There is a basic structure, which is the hydrogen atom consisting in a single proton with an electron in the ground state; but with all the heavier atoms more entities and processes are involved. Something of the degree of complexity is illustrated by the enormously wide variety of subatomic constituents: a recent count discloses forty or more different particles. Atoms are composed not only of protons, electrons and neutrons but also of positrons, mesons, muons, neutrinos, photons, and many others. Evidence is accumulating currently for the existence of a sub-atomic level, consisting of quarks, charm, gluons, etc., which furnishes additional displays of complexity.

(b) There is new strong evidence for the volatility of matter, which hitherto has been supposed to consist in only three states: solid, liquid and gaseous. Now an additional state of electrified gas has been discovered, the hot plasma state. This is the ionic, or excited, condition of matter, and it represents 95 percent of all the substance in the universe. (The other three are so rare that they have been described as 'trace contaminants'.) New vacuum techniques and microwave devices make it possible to study matter in the plasma state at high temperatures and low pressures where most electrons are no longer in their atomic quantum orbits.

The plasma state is defined by very simple laws outlining the electromagnetic interactions between nuclei and electrons, in the space between charged particles and magnetic fields, very common in the interior of stars as well as at the centers of galaxies (1). Most matter is too hot or too dilute; it is only at those special spots where quantum orbits can be formed that nature develops its atoms, its aggregates, its macro-molecules, and consequently its living organisms (2). Recently there has been

1 Victor F. Weisskopf, 'Physics in The Twentieth Century' in *Science*, 168, 923-930 (1970).
2 Victor F. Weisskopf, 'Physics in The Twentieth Century' in *Nature*, 168, 923-930 (1970).

evidence of a new range of plasma parameters in the earth's magnetosphere (1).

(c) Claims for the divisibility of matter have been supported by the list of atomic constituents. Discoveries of new particles are in fact frequently announced. Then, too, there are the energy-levels: there is the atom itself, and there is the nucleus with its short-range forces (2), and below that a supposed level of constituents of the elementary particles, an underlying structure of subparts currently classified as baryons, bosons and leptons, some of which were named above. And there is some reason to believe that the analysis into subparts does not stop there.

(d) Knowledge of the distribution of matter comes from the science of astronomy (3). Every successive estimation of the size of the universe revises the figures upward, and there is no apparent limit to the outer edges of the observed field; that is the story to date. Matter, then, is averagely distributed, according to observations made with X-ray telescopes which can go out to 200 million light years. The scattering gives a very low average density for the matter in metagalactic space (4). There is roughly as much matter between the stars as there is within them. This is an estimate based on visible masses. Recently evidence is accumulating that there is in addition a sizeable amount of invisible mass in the universe, dark matter for which only indirect methods of investigation exist (5).

(e) Studies on the distribution of matter in space disclose an overwhelming preponderance of volumes of empty space over volumes of matter. The characteristic feature of the universe, then, is not matter but empty space. Most of the interior of the atom is empty. Much the same thing can be said about the space occupied by the galaxies, to say nothing of the vast regions which separate them.

Matter as such is rare, and inert matter even rarer; it is not the vulgar stuff that previous centuries disdained but a precious substance in relatively short supply. Since we know that nothing can be done without at least a material base, matter is highly valuable. Bearing in mind the material range, from that of the motion of gross objects in the solid state to that of weak energy interactions, matter is an essential component of all activity.

1 Plasma is not only the prevailing state of matter in the universe, it is prevalent also closer to home. 'The earth's magnetosphere contains a rich variety of plasmas, from the relatively dense and cool ionospheric plasma to the extremely thin and hot plasmas at larger distances.' – Carl-Gunne Falthammar et al., 'The significance of magnetospheric research for progress in astrophysics' in *Nature*, 275, 185-188 (1978). For a more technical and extended account, see *Plasma Astrophysics*, ed. by D.B. Melrose, 2 vols. (London 1978, Gordon and Breach).
   D. Bohm, *Quantum Theory*, (Englewood Cliffs, N.J. 1951 Prentice-Hall).
3 W.T. Skilling, and R.S. Richardson, *Astronomy*. (New York 1947, Holt).
4 H. Shapley, *Of Stars and Men*. (Boston 1958, Beacon Press).
5 S.M. Faber and J.S. Gallagher, 'Masses and Mass-to-Light Ratios of Galaxies' in the *Annual Review of Astronomy and Astrophysics* (Palo Alto, Calif. 1979, Annual Reviews, Inc.), pp. 135-187, but see especially 182-3.

(f) The chemical analysis of the stars has been a much more limited affair, since it relies chiefly on spectrometer studies. The familiar elements, those ordinarily present on the surface of the earth, are the only ones detected thus far, but they are very common. This evidence lends some support to the contention that the matter of the stars, although in different chemicals compounds and in varying combinations, is very much the same as the matter of the earth. But although earth-like planets are probably frequent enough in many of the galaxies, planets and stars are no longer the only features of the cosmos. There seem to be a number of fixed types of galaxies as well as star clusters and occulting binaries, also vast hydrogen clouds and fields of ionic radiation.

Another piece of evidence from astronomy concerning uniformity comes from studies of the decay and death of the stars (1). Everywhere we look in the material universe we find the same entities and much the same processes at work. No single configuration of matter persists indefinitely but all is change and recombination. Due to the inter-convertibility of matter and energy, what recurs are the same structures, while an underlying substance remains capable of sustaining them.

3. The Kinds of Energy

Since, as I have pointed out, energy is substance in its dynamic state just as matter is substance in its static state, it was never a fixed stuff with which we were dealing but instead matter exerting its force.

I will be able to show the position of energy with respect to matter more clearly if I recall here some of the basic definitions which have been in use in physics and engineering.

Matter has been understood as the mass of a body, the numerical measure of its inertial force; and weight as the force of gravitational attraction of a body proportional to its mass. Force itself is that which changes or tends to change the state of rest or motion of a physical body, and is proportional to the rate of change in momentum.

Newton thought that force was external to matter. Since Newton it has been known that matter has the property of remaining at rest or in uniform motion unless acted on by an external force (Newton's First Law), and its acceleration has to be proportional to an applied force (his second law). Energy had to be separately explained when matter was thought to be irrefrangible. But now it is clear that energy is a property of matter and not merely something external to it, that matter itself can be transformed into energy and energy exerted as force. Matter, then, is potential energy.

1 M. Johnson, *Astronomy of Stellar Energy and Decay*. (New York 1959, Dover; O. Struve, *Stellar Evolution*. (Princeton, 1950, Princeton University Press).

In modern versions the weight of a body varies with its movement. The entity which remains unchanged during the changes in weight (the mass) is the inertial force. Since weight is the force of gravitational attraction, it produces an acceleration in a free body. Mass like force, has become a property of matter; it is the measure of that property in a particular body. Thus there has been a shift in mass from an irrefrangible static affair to one inevitably associated with its dynamical equivalent.

There are now known to be three kinds of energy at the physical level: nuclear, gravitational, and electromagnetic. Together they account for most physical activity. The nuclear forces are of two kinds, strong and weak; together they make up the organization of the atom. Electromagnetic forces account for most of the middle-scale phenomena, Coulomb forces in chemical reactions and compositions, for instance. Gravitational forces, at the other end of the spectrum, rule the stars and the galaxies.

However, our knowledge of energy transformations is far from complete. Consider the whole question of the existence of anti-matter for instance. Between matter and anti-matter there is a fundamental symmetry, for they have equal properties but opposite charges. Matter and anti-matter revert to pure energy in the reverse process of annihilation. The amount of energy concentrated in quasars cannot be accounted for on the basis of any familiar processes and so has yet to receive an adequate explanation.

As a result of modern physics the properties of matter are now known to exceed those which were formerly available only to common, unaided experience. A quick survey of the various departments of applied physics will furnish some idea of the wide variety and immense range now being explored and utilized. Electronics, thermodynamics, atomic and nuclear physics are only some of the chief subdivisions, and they themselves are also subdivided. Surely the end is not yet in sight, for such phenomena as superconductivity, semiconductors, ionized crystals, atomic fission and fusion, must be included. Clearly the properties of matter and energy are only beginning to be explored.

4. Material Particulars and Their Inseparable Properties

In this chapter thus far I have been considering matter *as such*. Now I shall have to turn to a consideration of matter as it occurs in particular things, and of course I shall always be counting among particular things not only atoms and trees and planets but also electric charges; in short, wave packets of energy of all sorts.

Since I have already offered a definition of matter, I shall now have to propose one for particulars. Accordingly, I define particulars as *material* individuals. I emphasize 'material' individuals for there is one other kind of individual, individual abstract objects, for instance, such as the individual number, '6'. Here we shall confine our attention specifically to *material* individuals. Throughout the remainder of this chapter, then, whenever I use the term, particular, it is the material individual that I mean.

Whenever we compare two particulars, we find in them both similarities and differences. There may be any number of similarities, we are not concerned with them at the moment; what we are concerned with is that there must be at least one difference. All individuals are members of classes, material individuals as well as abstract individuals: there are no absolutely unique things. But the material individuals differ from the abstract individuals in that each has at least one difference from all other members of the same class.

Particulars, or material individuals, can be distinguishable from abstract individuals by three inseparable properties which the former have and the latter have not. These properties are: (a) intolerance of opposition, (b) historicity, and (c) actual infinity. These cannot ever be found apart from the particulars in which they inhere.

(a) The first characterizing property of the particular is its intolerance of opposition, a feature which includes of course any active opposition. The particular resists physically, through the process of inertia any effort to make it other than it is. It makes the effort to remain itself and keeps in this way a certain integrity. To the extent to which a particular is organized, it has structural strength, and this supports its existence as whatever it is by offering resistance to change.

(b) The second characterizing property of the particular is its historicity. The particular endures through a succession of states in no two of which it is exactly the same, while in each state it carries over something from previous states; that is to say, it inherits its own unique past. By way of contrast, the abstract universal is always the same universal and nothing can alter it; but the particular, being a material individual, is changed in time: the cumulative feature of the particular is inherent in the fact that it *has* a history. The name for the particular when we deal with it in this context is 'fact'. A fact is a slot in the loose file of history.

(c) The third characterizing property of the particular is its actual infinity. The particular has its own peculiar kind of infinity. It is a product of history and so belongs to a past which nothing can ever change. It is what Whitehead called 'an objective, immortal fact'. For nothing that will ever happen can alter the fact that what happened has in fact happened.

The above three properties are inseparable from the particulars in which they inhere. One can hardly imagine a free-floating intolerance of opposition, historicity, or infinity, in the special senses assigned to each of these properties, for they are imbedded in the action and reaction which every material individual experiences at all times and places during the tenure of its existence.

It is important to note that each of these three properties of particulars stand in the way of the theory that particulars are nothing but bundles of universals. D.M. Armstrong has argued persuasively against that theory (1). If a particular is intolerant of

1 *Nominalism and Realism*, (London 1978, Cambridge University Press), 2 vols., vol. I, Part Three.

opposition, that means that it is stubbornly itself and resists any intrusion into its integrity. While it is true that universals are found in the particular as its properties and that the particular is the member of a class, that is not all that there is to the particular, because its resistance has a date and a place.

Each particular has a history, and this means a temporal sequence of interactions with other particulars which change it but not sufficiently to cost it its identity. The particular may be hedged about both above and below by universals, and that may be the reason for its investigation by scientists; but its material individuality remains. Usually there are detectable differences which mark the particular *as* a particular, but when all others fail there is always the distinction of date and place.

Experiments in science are always performed on material individuals (for this purpose we shall have to consider a finite group as an individual but not an infinite group). When a biologist in a laboratory dissects a rat, he is looking for properties it has in common with other members of the same class, for it is the class properties he is after. Still it is the particular rat that he dissects, he cannot dissect the class. The bedrock and basis of all scientific endeavor is the particular, the single material individual, whether it be a rat, or a more distant galaxy which can only be observed and measured.

That scientific investigation begins with the observation of particulars lends them a certain authority based on the assumption that they are real. I will describe the meaning of reality here as the utmost in reliability, and postpone a more technical definition. The success of the scientific method in making discoveries does provide some kind of endorsement of this leading assumption. Science has no lasting concern for particulars, it is looking for classes and their relations where the classes are those having particulars as members. If the particulars examined are typical, the information so obtained can be applied to other members of the same class, which is to say, to other particulars. All experiments must be repeatable if the findings obtained by means of them are to be accepted by the scientific community, and though repeating an experiment means employing a similar set of particulars, they are rarely the same ones that were engaged in the original experiment.

It is in general the properties of particulars which are studied, for the membership of the particulars in general classes is assumed. The scientists want to know what sorts of properties the members of classes can be counted on to possess. Thus the knowledge of certain kinds of universals is the aim of every investigation of particulars. The scientific method works upward from particulars, but begins — and this cannot be emphasized too strongly — with the assumption that the particulars are irrefrangable and have the properties named.

In recent decades applied scientists and technologists working in many of the major research laboratories have learned increasingly how to make new materials of all sorts having entirely new properties. New combinations of old materials, such as the metals, ceramics, and glasses, as well as entirely new materials, such as the polymers, are being improved on almost a daily basis. As the demand for them arises in science, engineering and industry, the 'materials scientists' are able to produce them. The range and flexibility of matter at this level of organization exceeds anything that was ever known or even imagined. The possibilities in this field of endeavor have only recently been discovered and so are nowhere near the end (1).

5. Material Particulars: The Separable Properties

In addition to the three properties of particulars or material individuals which are inseparable, there are other properties which are not so dependent and which can be separated, at least in the sense that they can be found in other particulars. These may be characterized as kinds of universals. Aristotle was at great pains to show that universals exist in material individuals and that is the sense of their reference here. In the next chapter I will discuss the different status of universals as they have their being apart from matter in a logical domain of their own.

The three kinds of universals or properties of material individuals which are actual universals are: (a) qualities, (b) forms, and (c) relations. A few words in explanation of each of these will be in order.

(a) A quality is that which is ultimately simple (2). Qualities as such are incapable of further analysis. Qualities appear only in particulars but are not integral parts of any particular in such a way that when that particular perishes its qualities as such perish with it. Every brown horse eventually dies but its color does not die with it, for the same brown appears in other horses and in many other kinds of material things; and wherever and whenever it appears it is the same color.

The topic of quality is important and will receive more extended treatment in another place (3). Here I might point to the special – and much neglected – quality of force (4). It is, like the other qualities, a sense property, in this case the sense of muscular pressure conveyed through the skin; but, unlike the others, it can change the state of a particular, for it is active. Most qualities are to be found only in some particulars, but force is a quality common to all, as the prevalence of gravitation shows. All existing particulars are subject to the forces which operate both in and on them.

(b) Forms are the structures of particulars. They are composed of internal relations.

1 Compare the state of materials science in 1967 when *Materials* by the editors of *Scientific American* (San Francisco, W.H. Freeman) was published, with the state now, as illustrated by the editors of *Science*, 208, 807-950, 1980.
2 *Foundations of Empiricism* (The Hague 1962, Martinus Nijhoff), p. 78.
3 See below, chapter IV, Section 2.
4 See below, chapter IV, Section 3.

Like qualities, forms do not perish when particulars do; instead they appear again in other particulars. Identical forms are common to classes of particulars: they have the peculiarity that they are self-contained in particulars yet link those particulars to others through resemblances. Forms are imperishable, and even when not appearing, always may appear. Moreover, forms are always the same and their appearance in a number of things is therefore also the same. The roundness of a round thing is the same form when it appears in something else round, just as we saw in the case of qualities that a special shade of blue is the same blue wherever and whenever it appears in a particular.

(c) Relations are the connections between particulars, and may be internal or external: internal when they connect parts in a whole, external when they connect wholes. The relations that connect particulars may be associated with any of them. 'To the left of', 'smaller than', 'fifth', are all examples of relations. Like forms, they exist among material things and also like forms they do not perish with material things: they may always recur.

THE INTEGRATIVE LEVELS, STRUCTURES AND FORCES

	MICROCOSMIC	MESOCOSMIC	MACROCOSMIC
CULTURAL	sign, artifact *persuasion*	ethical *value*	institutional *ethos*
PSYCHOLOGICAL	nerve *impulse*	cortical *awareness*	social *belief*
BIOLOGICAL	cell *genetic*	organic *life*	aggregative
CHEMICAL	molecule *Coulomb, valence*	elemental *reactive*	cosmic
PHYSICAL	atom *nuclear*	mechanical electro- magnetic	world gravitational

Table I
(read up; forces in italics)

6. Further Complications of Matter

The new understanding of matter, with its own set of conditions and its separable and inseparable properties, then, makes for an altogether different range of possibilities from those that were held to exist under the simple, rigid and restricted conception of matter of the older materialism which prevailed from Leucippus through Marx. What we refer to as 'matter' usually means matter with only the physical properties; but it is capable of sustaining many other properties as well, and we have better look at some of them even though they are not material in the strict meaning of the term but what could better be called 'material-supported'.

Perhaps the most remarkable property of matter is one that I have only touched on: its capacity for sustaining structures of high degrees of complexity and of developing in connection with those complex structures emerging qualities.

Atoms are capable of combining into groups. The resulting molecules make up the elements of crystals, gasses and liquids. Stones, planets, interstellar gasses, all of the gross material particulars perceptible by the unaided senses, consist in huge repetitions of comparatively simple chemical structures. The chemical substances are homogeneous collections of similar molecules.

The combining power which makes possible the next level of particulars is called valence. Further combinations result from mixed groups of substances having a bewildering number of properties. Elementary substances have atoms of only one kind; compound substances are made up of atoms of many different kinds. Already more than 400,000 chemical compounds have been catalogued and the end is not in sight.

We have been examining matter from the evidence furnished by the physical sciences: physics, astronomy and chemistry. As we continue our examination upward, moving from chemistry to the biological sciences, we find additional evidence of the same kind of matter but now supporting more elaborate forms and the emergent quality of life. We shall find that the matter of which the biological organisms are composed discloses much the same properties.

There is considerable evidence that life was produced spontaneously from inorganic materials (1). Protons and electrons combine into atoms, atoms combine into molecules, molecules combine into cells, cells combine into organisms, organisms combine into societies. the series is continuous and the dividing line between the inorganic and the organic a thin and perhaps non-existent one. Amino-acids have been produced in the laboratory by subjecting a mixture of gases such as might have been present in the atmosphere of the early earth to repeated spark discharges (2). It is known that proteins

---

1 A.I. Oparin, *The Origins of Life on the Earth*. (New York 1957, Academic Press); H. Shapley, *op. cit.*
2 S. Miller, in the *Journal of the American Chemical Society*, Vol. 77, 2351 (1955).

consist of long chains of amino-acid residues. Although cells behave in the body quite differently from what they do *in vitro*, this does not detract from the fact that all of the higher structures are the result of the complications of matter.

From biochemistry and biology it has been possible to learn of the deep affinity to matter of all living organisms. All intricate but continuous series of steps lead from the molecule to the cell. Cells are enormously complex structures of giant molecules, while protein molecules have been synthesized from less organized materials and their patterns discovered. It would seem that organisms are enormously complex chemical compounds even more complexly organized and integrated. They are not generically different even though they may be structurally higher.

At the level of human society another factor begins to appear. Materials in the environment are altered to suit human uses, and we see for the first time the construction of an artificial environment and of man's attempt to adapt to it. For as I have noted many times in this book already, the fact is that it makes no sense to talk about man without including his tools. Biological man has never been without them, and they have become an increasingly crucial factor in his development. He cannot be considered without his technology.

It is customary to pursue the analysis of matter past the biological organisms. But man leads off into a divergent but higher set of structures called societies, which are organizations of men and technologically-produced artifacts together with their institutions, their cultures and finally their civilizations. Civilizations, containing all of the achievements in the arts and sciences, certainly do represent the highest attainments of matter thus far.

Animals, including the human, evolved from lower forms, and some of them are still evolving. Animals, historically speaking, ride to their adventures: death for the dinosaur after a life for the species of some 140 million years, four million years of life unchanged for the still flourishing horseshoe crab, and evolution for the ancestors of man.

Regardless of whether there is or is not life on other planets in distant galaxies (and the preponderance of evidence seems to indicate that there may be), we have to account for its appearance here. One geologist has speculated that the evolution of the earth itself, and in particular plate tectonics, may have been responsible for the emergence of man (1).

The universe is large, and the conditions found on earth must have occurred many times. If life arose spontaneously from inorganic materials, then it could exist wherever there is the same set of conditions. If Darwinian evolutions holds, there is no reason to suppose that the evolutionary process which produced man necessarily always stopped there. Other planets might have been the scene for the development of species higher than man, provided only that life there has gone on somewhat longer.

1   S. Miller, in the *Journal of the American Chemical Society*, Vol. 77, 2352 (1955).

7. The Soft Variables

We are now in a position to conclude by saying a little more about matter. The properties of matter were once held to be few and known, now they are understood to be many and largely unknown. This weakens the assumptions and widens the implications.

In what it may be possible to call the new materialism, matter can support all of the properties of which we have knowledge, whether these be substantial or logical. The richness of matter renders it capable of containing more than traditionally has been attributed to it. Matter is no longer a term confined to familiar objects. It has been shown to be extended in two directions and applies to objects accessible only through instruments and mathematical calculations: microcosmic objects, such as atoms and cells; and macrocosmic objects, such as star clusters and eclipsing binaries. And so we are no longer confined to the knowledge and experience of objects in what now may be described as the mesocosmic range. Although there has not yet been a successful penetration inside the electron or outside the metagalaxy, it is clear that we live in the middle segment of a three-segmented universe. But this topic, which belongs more properly to cosmology, will be examined below in chapter V.

It is our last task, here, then to show how certain other properties which had been previously excluded from matter can now be shown to be included.

What are these other properties? They could best be characterized perhaps as soft variables: those which arise from matter when it is highly organized in a complex state. 'Mind', 'consciousness', 'spirit', 'purpose', 'goodness', and 'beauty', will serve as typical terms.

Let us begin with 'mind'. In the general use of the term there is some ambiguity. Descartes, for instance who proposed an ontological dualism of mind and matter meant by mind everything not covered by matter: thoughts, certainly, since the term for mental things was *res cogitans*; but probably also consciousness, and an unnamed and undefined spiritual property. Certainly, the term 'mind' usually includes memory and thought: the retention and manipulation of logical entities and their combination.

In the traditional materialism, there was no explanation for mental events. They were not supposed to be material and so were assigned to a separate though often parallel series without interaction, as for instance by Malebranche. In the newer materialism, mental events take place in the brain, and consist in signals and signalling systems both of which are at the very least material. Experiments with the electroencephalogram, with drugs, and with ablation procedures, have indicated that an intimate relation exists between mind and brain.

Investigations are now at the threshold of understanding of the relations between language and the development of the brain. Memory plays a role here as pathways are laid down; the use of language depends upon memorizing the ways in which it is used.

The technology of meaning, reference and communication is intimately involved (1).

'Consciousness' would seem to depend upon a particular area of the brain, namely, the midbrain section of the upper brain stem reticular formation (2). Consciousness is a state of alertness; to remain conscious is to be aware of some segment of the external world or of its representation, and today that means for most of those in western countries an artificial environment or its mental effects.

'Spirit' is a more difficult concept to pin down. Its use has been varied. Among the meanings which may be distinguished are: the animating of the vital principle of man, the soul, immaterial being, the unknown.

If we understand by spirit the dominant inner quality of a material thing, then in man it is his highest function, the direction of consciousness in terms of attitudes and aggressions. And if we assume that the material universe as a whole has its own dominant inner quality, then the religious enterprise is intended to put man in touch with it. What is called the life of the spirit is the effort of man to exceed himself through feeling by recognizing in that feeling a similarity with the quality of the greatest whole. Religion is man seeking identification with the material universe or its cause. The defender of spirit, then, can accept the findings of the materialist. The spiritualist believes in matter, he simply thinks that there is more to it than formerly had been supposed.

If being is defined as material things and forces together with logic as their representation *in absentia*, then 'immaterial being' becomes a contradiction in terms. Presumably the meaning is the same as the one I have just been discussing, that some of the complex properties of matter had not customarily been associated with it but mistakenly held to be opposed to it. And so the conclusion, given the premises of materialism and its rules of admissible evidence, will be the same, namely, that spirit is a property of matter.

'Spirit' as the unknown is easy to understand but difficult to dismiss, for there is a marked difference in the attitudes toward it of the materialist and the idealist. The materialist finds an inexplicable mystery in the known, the idealist finds it only in the unknown. Moreover, the materialist wishes to penetrate farther into the unknown and by means of technology and artifacts reduce it to the known, for it is knowledge that he seeks. The idealist, on the other hand, would leave the unknown alone, and only wishes to stand in awe before it. At the same time he holds such responses to be a kind of superior knowledge.

The difference between the character and extent of the knowledge of primitive man and civilized man is remarkable. When we compare their religions we find that each claims avenues of knowledge of a super-natural and trans-sensory variety which authorize

1 See further on mind, chapters VIII and IX, below; on communication, see chapter XII, below.
2 H.W. Magoun, *The Waking Brain*. (Springfield, Ill., 1958. Thomas).

actions on matter however violent. The consistency of the materialist is that he would use his knowledge of matter to alter material situations. The inconsistency of the idealist is that he would use his knowledge of spirit in the same way (1).

Next we have to examine the property of 'purpose'. For individual man it means quite simply his pursuit of a goal. In material terms, this could be the service of society arranged in terms of material culture, with its *terminus ad quem* unknown. For society itself it is necessary to appeal to the principle of biological evolution. *Homo sapiens* is working itself out to an end which lies invisible in the greater expanses of the material universe. But we do know this much. A state of maximum entropy would seem to be the *terminus ad quem* of the universe, with a state of maximum evolution bringing it up somewhat short but consisting in the perfection of the human successor animal (2).

The last properties we shall have to consider are those of 'goodness' and 'beauty'. They will be briefly considered here because both are properties of matter, but they will receive extended treatment in later chapters, X and XI, and XIV, respectively.

That 'goodness' could be considered a property of matter is not a currently conventional proposition, yet it is one which can be defended. Let us define the good as the quality which emerges from the attraction between material objects. It would not be too difficult to show that all material objects are so related. The qualities change in accordance with the complexity of structure of the objects. Thus, to give two disparate examples, the quality of goodness at the level of the physical is the pull of gravitation, at the psychological level it is the bond of friendship or love. The good, in the conventional sense, would be the attraction between human individuals. The increase of good can be accomplished by resorting to artifacts. Technology is the resource for the increasing of goods (3).

'Beauty' as a property of matter is perhaps more unfamiliar. Artists have been proclaiming the beauties of nature for millennia. Beauty can be defined as the quality which emerges from the perfect relations between the parts of a material object. Beauty is thus an internal property. Art in the conventional aesthetic sense is the quality of beauty produced in a material object through human agency. It does not belong in the eye of the beholder, which would otherwise see everything the same, but is a quality of the object: a sunset, a tree, a woman's body. Beauty is objectified and increased in works of art, which are aesthetic artifacts produced by a material technology specifically designed for that purpose (1).

That ethical and aesthetic values can be properties of matter seems strange only because the spiritual has been separated from the material; and while it has been freely acknowledged that the human individual has a spiritual nature, the ethical and the aes-

1 For further on religion, see chapter XV, below.
2 For further on purpose, see chapter XVI, below.
3 See further on morality, chapters X and XI, below.

thetic have been considered to belong to the spiritual. Then, too, they have been considered exclusively human. But there are reasons for supposing that both goodness and beauty must be properties of matter one of which is that they are qualitative.

All sort of perspectives are opened up by this new conception, but with whatever analytical or integrative levels we are dealing, it is still the same matter, a static substance having as its property a potentiality of reaction. The unity of the universe is a material unity. It can be summed up somewhat as follows.

THE VISIBLE WORLD

The contents of the integrative levels in disorder.
The properties of matter; energy-bundles, qualities.
The state of matter, gas, liquid, solid, plasma.
Space: the extent of occupancy.

Table II

---

1 See further on aesthetics, chapter XIV, below.

CHAPTER III

UNIVERSALS: THE FORMS

It is not often sufficiently recognized that there are two kinds of universals: forms and qualities. In this chapter I will treat of the first kind, and in the next of the second. Pursuant with the purpose of showing the influence of science and technology in every department of philosophy, I propose to begin by looking at the scientific evidence for the reality of universals and only then to ask what philosophical theory it supports. What is chiefly of importance here is the dependence of the scientific method of investigation on the development of the appropriate technology and the production of its special variety of artifacts: a scientific laboratory, an astronomical observatory — what are these of not large collections of complex and ingenious artifacts, the products of an elaborate technology? Before we are done I hope to show that there is a connection between the scientific method and the argument for the reality of universals, and that philosophy can benefit from exploring that connection.

1. Material and Logical Objects

Work in the sciences has disclosed that there are two distinctly different kinds of objects with which the investigator is engaged, material and logical. Experiments deal with material objects: with things and forces, mathematics deals with logical objects, such as numbers. Ordinarily, both sorts of objects are referred to by means of names.

All words are names, and for all names there are the objects named, things which must have existed before they were named, a position which assigns the ontological priority to the material things.

There are four and only four kinds of things. All words name one of the four kinds: (a) material things, (b) properties of material things, (c) classes of material things, and (d) classes of classes of material things.

Examples are: material things, such as the White House in Washington, D.C.; properties of material things, such as 'round' or 'blue'; classes of material things, such as 'horse' or 'planet'; and finally classes of classes of material things, such as 'include' or 'six'. A few extended comments on each will be helpful.

(a) The term, material things, needs no extended explanation here. I have already defined matter as static substance and substance as the irrational ground of individual reaction. Matter and energy as we noted are interchangeable, and so we must include under the heading of matter all material objects, such as forces, force fields, energy-bundles, etc. We can now understand matter and energy as forms of substance.

(b) Properties of material things are better known in traditional philosophy as qualities and relations. Such properties must be distinguished from proper parts, which are themselves material things, for instance the leg of a chair is a material part of the chair. Properties of material things, then, are not strictly speaking material things except in a special sense; they are rather to be classified as kinds of classes.

(c) It is often difficult to recognize that classes of material things are not themselves material things but classes, and classes as such are abstract. The class 'tree' is abstract; it is not a material thing, like a tree. Classes of material things have to be distinguished from that other abstract class, the class of logical things. The class 'three' is abstract, but it is a different order of abstraction from 'tree'. Classes of logical things are more abstract than classes of material things.

(d) Classes of classes are classes whose members are classes, and then of course there are classes of classes of classes, etc. In this category, as we shall see, are to be found all of the concepts of pure mathematics, all numbers, for instance.

It comes down to this, that of the four objects named, only one is concrete, the other three are abstract and formal – in a word, universals. Properties of material things, classes of material things, and classes of classes of material things, are all universals. Only the material things themselves are concrete. There is something very special which needs saying about each of them.

First as to the material things. Let us consider those of a given class, say planets or human individuals, it does not make any difference which we choose. For the point made before must be emphasized here: that most material things that exist are absent.

The vast preponderance of material things of a given class are absent in time as well as in space. The understanding of universals has been profoundly affected by the knowledge of astronomy. Planets are absent in time because they last for a long while even if not forever. During the 13 billion years or more that the material universe has existed, many planets have come and gone. They must be counted as members of the same class even though the are not present. We encounter only those in our neighborhood and regard them as typical. But given the new information concerning the size of the cosmic universe, it may happen that the number of planets is infinite. Out as far as 200 million light years, which now appears to be the radio horizon, astronomers encounter similar configurations of galaxies. And given that each galaxy contains billions of stars, the number of material objects of the classes we have chosen could well be infinite.

It is possible to say of the class when its membership is denumerable that it is extensively finite, but we are more concerned here with the class which is extensively infinite, but it is not possible to say of it when its membership is non-denumerable that it is extensively infinite even though it may be. A class of material objects does not give us a clue as to the number of its absent members. For that, we have to go to logical objects, which do. How many planets are there in the material unvierse? We do not know, but we do know that the number of circles is infinite because the class 'circle' is inexhaustible. But it is possible to say in any case that one of the conditions sets for the observer by the encounter with material objects is that he is confronted by a class. Confrontation in sense experience means the encounter through the present members of a class with that class.

So we shall see that the encounter with a logical class, i.e. one whose members are classes, is no less a confrontation, and we shall be led to repeat the statement made at the outset, namely, that there are two kinds of objects, material objects and logical objects. A logical object is a class which has as its members either properties of material objects (forms or qualities), classes of material objects, or classes of classes of material objects. One of the unique features of this analysis is the classification of qualities as one kind of logical object.

However, we will be concerned for the moment exclusively with the class whose members are material objects. The yellow moon is present to the senses of the observer because they are passive in the encounter and conditioned by the experience. Not only is the moon there and now, it is also the member of an indefinitely large class. Its presence can be considered from two aspects, the one particular and the other general.

The important thing to note about the particular aspect is that it is capable of calling upon all the observer's powers of apprehension though not necessarily all at one time. To begin with he sees it, then if it were any material object closer than the moon he could if he liked touch it. But that is not all of the experience either of sight or touch, for if he could walk around it he would be able to see and touch other sides of it. This fact brings into play the dynamic phase of the encounter referred to earlier. For it is only as an entire man that the observer is able to plan and execute all of his experiences.

What does this do to our conception of classes of material things? It makes them possibly infinite. Concrete objects, in other words, may be as infinite in number as the abstract objects which were always known to be infinite. There is no limit to the number of round or blue material things. The only difference between abstract objects and concrete objects – between, that is to say, universals and classes of material things – is the question of the extent of cover reference.

There are evidently two kinds of universals; formal or abstract universals, the objects of mathematics, and material universals, the objects corresponding to our sense experience. And we must now acknowledge that they may be equally infinite.

The special word to say about abstract universals is that they are not classes at all in the logical meaning of the term; instead they are iterations. An iteration may be defined as a repeated identity, and as such it differs from the older conception of the universal. It resembles the older universal in that it can occur in an indefinitely large number of instances, being both inexhaustible and undecidable. It resembles the particular in that it can be sensed, and yet every instance of its actualization is like every other. This is no doubt what Plato had in mind when he said of 'the ideas or forms that each in itself is one, but that by virtue of their communion with actions and bodies and with one another they present themselves everywhere, each as a multiplicity of aspects' (1).

The abstract universals are iterations, while the material things are individuals and so unique. This is the most basic distinction in philosophy to be supported by science. There is one and only one triangle, although it has many appearances; but there are many cats, each one a separate and distinct individual as well as the member of a class. The triangle has been defined as three points not on a straight line; it always has three, and only three, sides and angles. But the particular cat may have a black ear, a short tail, and other characteristics not found in most other cats, for 'all cats are feline' but not 'all cats are black'.

No relation holds between a term and itself, and so an identity is not a relation. Iterations are in a sense identities but they are not relations. For when we say that 'x is x', if we are talking about two different things there can be no identity, only an iteration; and if we are talking about the same thing twice, there cannot be one, either. We must distinguish between a form and its expression, since to say that x is identical with itself is meaningless. One circle may be larger than another circle but *pro forma* they are the same circle. The abstract objects, then, are logical constants, in the sense that they are iterations rather than classes having members. '2' is a logical constant because it has no members, only instances.

2. Propositions

From classes we now turn to propositions. Propositions as such connect classes. When they do so they form new classes, but this time with a difference from the way in which entities are named, for names are arbitrarily assigned to classes and can be false only when incorrectly assigned. Frege introduced the distinction for propositions between meaning and reference, and the references of propositions are not always cor-

1 *Republic*, Book V, 476A. (Paul Shorey, trans.).

rect. The incorrect references are those of false propositions. Of course the meaning of propositions also can be false, as when a logical object is incorrectly defined or described. The meaning of the proposition 'Three is an even number' is false.

There are three kinds of propositions: singular material propositions, universal material propositions and universal formal propositions.

The scientific operation begins with observations and experiments and then rises inductively to empirical generalizations. The observations and experiments clearly involve material individuals, and are reported in terms of *singular material propositions*. The inductive generalizations are reported in terms of hypotheses, which are *universal material propositions*. In pure mathematics the situation is simpler: all mathematical equations are in effect *universal formal propositions*.

From singular material propositions to universal material propositions to universal formal propositions, there is an increase in the degree of generality, marked by references to increasingly larger areas of spatial occupancy. Let us look at the difference more closely.

A 'singular material proposition' is one which refers to an individual material thing. A 'universal material proposition' is one which refers to all material things of a given class throughout the cosmos. A 'universal formal proposition' is one which refers to everything in the cosmos without restriction.

Attention is called to the fact that it is to all the material things of a given class – the class in extension – and not to the class itself that the universal material proposition refers. Both singular and universal material propositions refer to concrete things, for no difference exists in this respect between 'one' and 'all' and only the question of quantity is involved.

The singular material proposition refers to an individual material thing occupying a limited portion of space. Thus we say, 'Abraham Lincoln died in Washington in 1865'. Language becomes extremely cramped when we need to refer to unique material things, because language itself is inherently general. It fits abstract things better than concrete things. Except for the attribution of properties of classes, both of which are abstract, language as such was not designed to refer to individuals beyond their naming and locating in space and time, which is a kind of pointing *in absentia*. Ostensive reference is employed so seldom after the initial identification as to constitute a trivial instance. The moment we wish to add anything to a name or to qualify it in any way we perforce resort to the use of the names of abstract things which we then proceed to put in some equally abstract relation with each other.

A 'universal material proposition' is one which refers to all material things of a given class throughout the cosmos. If I say, 'All men are mortal', I refer to all men no matter when or where they live, always and everywhere if need be. This may take us into an indefinitely long past or an indefinitely long future (or both), and throughout the material universe, as well as here and now on the surface of the earth. But even so, not all of space is named in the reference, only those spaces, however many, which contain mortal men. The proposition refers, in other words, to a particular kind of spatial occupancy, namely, the occupancy of space by men who are mortal. Since space can be occupied only by matter; reference to spatial occupancy or to material things does not differ materially.

To say that a universal material proposition refers to all material things of a given class throughout the cosmos is not to say that there is nothing to a universal material proposition except material things. For in addition to the material things to which the universal material proposition refers, and even in addition to all of them, there is the proposition itself. A proposition is a whole, and with propositions as with most wholes the whole is more than the sum of its parts. That 'more' is its formal structure. Matter (or energy) always occurs in some form. Formal materialism has been much maligned in the name of materialism.

Information is recognized and stored in linguistic form, a most fortunate arrangement when we consider that both knowledge and language are incurably general. Ordinarily, it is supposed that because the reference of singular propositions is azygous they must be peculiarly local, but a little examination will show that this need not be the case. For there is a sense in which what I have said of the *universal* material proposition is true *mutatis mutandis* of the *singular* material proposition. 'Abraham Lincoln was assassinated' means that for every Abraham Lincoln in the universe this was the manner of his death. 'Lyndon B. Johnson was President of the United States' means that for every Lyndon B. Johnson and for every United States in the universe this has to be true. Given the high probability of the prevalence of life on other planets throughout the universe, as indicated by recent astronomy, we have no warrant to assert that either of the above two propositions refers to unique occurences. Thus in the last analysis what is meant by 'unique' is 'being the only existing member of a class which may (or may not) contain more than one member'.

A 'universal formal proposition' will be one which refers *indirectly* to all material things without restriction as to class by referring *directly* to abstract things. Universal formal propositions are therefore abstract locial propositions. They hold throughout the cosmos because they refer to all of it, and to no particular portion more than to any other. They constitute, in other words, reference to the elements of a separate domain. If universal material propositions refer to all of occupied space by referring to material things of a certain sort everywhere, universal formal propositions have a broader reference because they are on a different basis. Universal *material* propositions refer to material things of a certain sort and hence to limited portions of space, namely, to the spaces which contain those sorts of material things.

Universal *formal* propositions do not refer to any particular class of material things and hence are not limited as to space. Thus they are the only propositions which refer to *all* of occupied space. They have, as it were, unlimited extension. What holds for material things must hold for the propositions which refer to them because of the correspondence between the propositions and the material things; whereas what holds for formal conditions does not require either particular classes of material things or particular spaces, but does hold for them everywhere.

3. The Scientific Method

Now we can address ourselves to the question of the evidence for universals in science and technology. Full abstraction discloses that there is only one model for the hypothetico-experimental laboratory procedure of the physical sciences. Although its employment in the separate physical sciences is always mediated by extenuating circumstances, it is still true that essentially the same set of procedures, conducted in approximately the same order (though often by different sets of investigators and on different occasions), can be found in laboratory practice.

The scientific method is a continuing process, which nevertheless lends itself to analysis into seven well-defined stages. These are: *observation, induction, hypothesis, experiment, calculation, prediction* and *control*. Each of these stages except the first emerges logically from the one before, and each except the last leads logically to the next.

Observations usually made by means of instruments are first conducted in order to uncover provocative facts. What are observed are of course particulars: material individuals, and this is the beginning stage of purely descriptive knowledge. *Inductions* from such facts are made next in order to discover *hypotheses* worthy of investigation. This is the stage of brilliant originative insights. The hypotheses are set up for testing in this fashion, and the tests are then carried out in three ways. The first way is the one peculiar to science; it involves the checking of hypotheses by means of *experiments* which again involve complex collections of instruments. In the second way the hypotheses are matched against existing theories by means of mathematical *calculations*. This is the stage in which quantitative laws are shown to be the necessary logical consequences of a few axioms or assumptions. Finally, the third way is to make predictions from the hypothesis and to use them to exercise *control* over practice. Those hypotheses which pass all three tests successfully are considered to be established, however tentatively, as laws.

A word of reservation: not every working scientist engages in all seven of these activities, but the sequence is somehow maintained. The multi-stage process is a description of the overall method. It can be, and usually is, repeated again and again, for it is self-corrective. There is of course nothing automatic about it. Like all logical structures, it needs to engage the intuition and constitutes an aid to it, not a substitute for it. We do not understand the method completely unless we include the role of intuition at every stage.

The scientific method is a method of discovery. What is it, exactly, that is discovered? A rough survey discloses that there are eight different kinds of knowledge which can and often do result from the application – or repeated application – of the method. These are: *empirical systems, empirical areas, entities, laws, processes, formulas, rules* and *procedural principles*. I plan to look more closely at only two of these: entities and laws.

Entities are classes of particulars, such as 'electrons' or 'cells'. Experiments are always performed on single entities regarded as typical members of those classes; and so in the end entities represent classes even though they are classes of a special sort, that is to say, classes having material individuals as members.

Laws are combinations of universal classes. Finding a law in an empirical area is like inserting a knife where there is a natural joint. The knowledge of laws, though not of course the laws themselves, is the result of repeatable experiences. 'Empirical laws' may be called laws, but they are usually summaries of the tendencies of phenomena. That heavy industry tends to pollute the atmosphere, is an empirical law. Similarly, statistical laws are generalizations derived from the study of large populations of instances.

Scientific laws are descriptions of forces at work in nature but are not themselves those forces. The law of gravitation is not known to have any exception and therefore gives every indication of being a universal law. Yet it is neither a mere summary nor a cause: the law of gravitation does not *make* bodies attract each other: they do so if they are bodies, and it is the regularity of their behavior that we have come to call the law of gravitation. The chief point however – often overlooked – is that the regularity is independent of the bodies.

We seem to be dealing with something above and beyond the phenomena, for neither the fallacy of composition nor that of division can be allowed to be applied here; the properties of gravitational instances of material bodies exercising attraction are *not* the properties of the law of gravitation, and, conversely, the properties of the law of gravitation are *not* those of material bodies attracting each other. There are two separate and distinct, though quite intimately related, levels, and the relationship is one between two natural domains.

The hypothetico-experimental method of investigation exists in order that scientists can discover laws by means of which it is possible to predict and even sometimes control the behavior of bodies in the material world. The method calls for laws to be discovered among facts, and also for the independence from facts to those laws.

The laws themselves are members of a class of universals. They are not the only universals, of course; pure mathematics is composed entirely of them; but they are authentic universals and they refute forever the contention that universals are not real. Not that all universals are real of course, for then every common noun would represent a real universal. We are more interested in learning *which* universals are real, and the scientific method was devised to do just that.

4. The Pure Universals

The whole conception of universals independent of space, time and matter, as established in traditional philosophy since Plato for some time now has been vigorously rejected by empirically-minded investigators. They have understood universals, however, chiefly as ideas in the human mind. Modern advocates have not made the best case that could have been made for them. Philosophical concepts are best used when we do not think *about* them but *by means* of them. It would be difficult for anyone working in pure mathematics to reject the conception once he had understood it in his own professional terms, and it would be equally difficult for an experimental scientist who sought the laws in his own field. There are many sorts of universals, as I hope to show, and the knowledge of them, often hard to come by, is the aim of every objective investigator.

Universals as perfectly abstract entities occupy a domain of their own where they have their own meanings and relations. The best example of pure universals must come from mathematics. We might look into this first before discussing the role of universals in science and technology.

Current opinion in mathematics bases it on set theory, and set theory in turn rests on sets and their members. A set is a collection of objects, called the members of the set, which completely determine it. Set theory however begins with the empty set, the set which has no members. The foundations of set theory, someone has observed, are firmly seated in the air. For the empty set has no reference to any content: no material individuals, nothing. It is as nearly abstract as any item can be.

Beginning with the empty set, there is next the set whose only member is the empty set, then the set consisting in the empty set and the set whose only member is the empty set, and so on. Everything else can be built up from there, so that with the aid of axioms and the theorems deduced from them, it is possible to arrive eventually at ordinal numbers. Cardinals are ordinals; an ordinal is the cardinal x if it is the least ordinal equinumerous with the set x.

Cardinals are of many kinds. The most familiar of course are the 'natural numbers' or positive integers, 1, 2, 3, ..., but there are many others such as the infinite numbers, the transfinite numbers, the transcendental numbers.

The domain of mathematics if we count the untold richness of its three broad subdivisions: algebra, analysis and topology, is everywhere dense and has only begun to be explored. But everything in it is independent of the domain of matter in space and time. A triangle is no different now from what it was in Euclid's day. It would still be possible to describe a triangle as Euclid did in Book I of *The Elements*, as a trilateral figure enclosed by three straight lines.

Mathematics is the ideal of universality at which all general statements aim. Mathematical laws are logical connections, they are of the form 'If A, then B'. To say that mathematical references are to *nothing* in particular is also to say, paradoxically, that they refer to *everything* in particular. Of any three material items in the cosmos, it is true that they are three and an instance of '3'. There are not, indeed there cannot be, any exceptions.

This is not to say, however, that the knowledge of mathematics was not first suggested by relations prevailing in the material world. Consider its most primitive formulation, the set and its elements; was not this an abstraction from the resemblances of material things, from, say, the observation that horses resemble each other enough to suggest the notion of a class of horses? The abstraction of mathematical concepts from the material world must be regarded, however, as an act of discovery rather than one of invention. The concepts of mathematics have their own relations whose character could not have been anticipated and therefore whose properties are their own.

The recognition of the abstract condition of mathematics and its power is evidenced by the use of the language of mathematics in the experimental physical sciences. Their laws are not equally abstract since their expressions have a substantive content, but they are stated in mathematical terms as much as possible. Scientific laws are laws framed in accordance with mathematical formulations. The result, as we will shortly see, is another order of abstractions, one which is general but never absolutely so, never quite attaining to the perfect generality of mathematics.

The technological aspect of mathematics is of course represented by its language, especially in writing and printing. Peirce insisted that mathematics is an observational science, and his evidence was its use of graphs. It is often possible to plot some information on a graph, which enables the observer to read off others. But in any case without the use of special symbols mathematics would have remained unknown.

5. Universals in Science and Mathematics

The most important conclusion from the foregoing is that both the laws of science and the equations of mathematics are universals. The laws of science are universal material propositions, while the equations of mathematics are universal formal propositions.

The laws of science may be classified as universal material propositions; that is to say, they are stated abstractly yet retain a concrete reference. The aim of all scientific investigations is to conduct further tests of hypotheses in order to determine whether they can be framed in terms of the universal formal propositions of mathematics. They are intermediate between the formal and material domains, having something of the character of universal formal propositions and something also of a material content like that of individual particulars.

Ontologically speaking, they are very peculiar affairs, for they partake of the generality of abstractions but they do so without losing the particularity of material objects. They describe accurately the regularities prevailing in the materal world, but in abstract terms. Consider for example Newton's inverse square law of gravitation: it is stated for a vacuum but it gives the conditions of material individuals without exception. A scientific law cannot be reduced to the objects to which it applies: the law of gravitation is not the sum of bodies attracting each other; but neither can it be listed among the abstractions of pure mathematics. It belongs oddly enough somewhere between logic and matter.

Universal formal propositions are the propositions of logic and mathematics. They refer to all of space without exception because they do not refer to any space in particular but only to the formal conditions under which space can be occupied, conditions which can be met by any occupancy of space. Universal material propositions refer to any space where certain specified material things may be found to exist, but universal formal propositions require no such specification.

In the words of Professor Warren Goldfarb,

'logic carries with it the notion that logical truths are completely general, not in the sense of being the most general truths about logical furniture, but rather in the sense of having no subject matter in particular, of talking of no entities or sorts of entities in particular, and of being applicable no matter what things we wish to investigate (1).'

This, then, is the one distinction between universal material propositions and universal formal propositions, that the latter involve the whole of spatial occupancy rather than a part, such as the spatial occupancy of all material things of a given class. The reference of formal propositions is to all material things of all classes with respect to a given property and hence to all of the space which they might occupy.

An illustration might perhaps make this somewhat more evident. If I say, 'Water is fluid and transparent', I am talking about some of the properties of a particular kind of matter in those limited portions of space which contain it. But if I say '2+2=4' I am talking about some of the properties of matter without regard to a particular kind and indeed not limited by any restrictions as to the properties which particular kinds of matter must have, so that I am referring to conditions which hold for all of spatial occupancy.

There is no way to check on the correspondence of universal formal propositions. Instead, we try to construct support for them by building them into a system in which by following rules of inference from propositions intuitively suspected to be true, they furnish the basis for the deducibility of other propositions. But the assignment of universal formal propositions to the abstract structures of logic and mathematics in which

1 Warren D. Goldfarb, 'Logic In The Twenties: The Nature of The Quantifier' in *Journal of Symbolic Logic*. 44, 353 (1978).

they appear as axioms or theorems does not get rid of their correspondence. And the fact that what they correspond to is as universal as all of space should not be too surprising.

Universal material propositions refer directly to the occupancy of space by matter, universal formal propositions do so *in*directly. If there is a correspondence between a proposition and the objects to which it refers, this can be absolute if the things are unrestrictedly general but not if they are not. Hence universal formal propositions refer to general things, abstract things. Instead of referring directly to all material things in the whole of space, they refer directly to abstract things, that is to say, to the class. We *allude* to the class of all formal reference, but we *use* the class because of its convenience as a quick route to extension. If we say, 'the class, horse', we do so because in this shorthand way we can refer to every animal of the class. And if we say 'two', the reference to this class is wider still because it refers to all pairs whatever they may happen to be, horses, or Presidents of the United States, or the like.

The point is that referring to an abstract thing is the same as referring to all of space with respect to a special kind of occupancy. Direct reference to a class involves indirect reference to all of space with respect to the existence or absence of the property indicated by the class. Abstract classes name the peculiar features of universal spatial occupancy. The difference between referring to all the members of a class and referring to the class itself is not lost simply because all of the *members* of a class are members of a *class*.

Perhaps the term, 'field', made familiar by modern physics will be of help here in the understanding of reference to space. An electric or magnetic field is the area in which a force is exerted. Consider space as the field and the functions of universals transferred to those of the field. Then the formal universal is a material reference.

That classes are pivotal reference points between universal propositions and material things in no way detracts from their reality. Class-membership in the distributive sense, which takes account only of material things, does not preclude the reality of class-membership in the collective sense, which takes account of the reality of classes. The similarities between material things (including their qualities and relations as well as membership in classes) have a permanence that the changing material things themselves do not have. When we relate classes to each other and when we explore classes of classes, we behave toward these abstractions as though we assumed their independence of the concrete world. They are independent because they have their own domain in which rules which are peculiar to them apply, such as the rules of inference, and also because rules which apply to the concrete world apply to abstractions with more rigor, such as opposition or conflict in the concrete world and contradiction in the abstract.

All propositions and not merely the universal variety belong to the abstract world governed by logic. All material things and the spaces which they occupy belong to the concrete world of substantial reaction. References connect them. It is possible to claim, then, that the abstract world is the second tier of a two-tiered world in which the knowledge of abstract things has to be gained through an examination of concrete things, for the abstract things hold sway over the concrete things but never completely so.

6. The Platonic Conception

How does the foregoing conception of universals fit with the traditional one? Ever since Plato, philosophers have been debating the status of universals. Let us turn back briefly, then, in order to compare the new conception with the one advocated by Plato.

Plato's interest was in the intelligible order, and he rarely let it descend to the visible world because he was convinced that nothing could be learned from the objects of sense but only from pure ideas (1). He went directly to the abstractions, treating them as though they had not been abstracted. I suggest that the way of empiricism is slower but more reliable. As empiricists we shall have to reverse this procedure and, taking both the intelligible order and the visible world seriously, move only on cue from the visible world. Besides, as we now know, there are things in that world that were never dreamed of in Plato's philosophy, configurations that can only be observed by means of instruments, and regularities that we now must recognize as being among the elements of his intelligible order.

There was nothing in Plato's day that could have served as a criterion, and that is why he took the theory as far as he could, and if it was not far enough, we must remember that it was as far as anyone could have gone with it then. Fortunately, the situation is no longer the same. We still have the lines drawn more or less as he drew them: between the intelligible order, even though no longer presided over by the good; and the visible world, even though no longer ruled by the sun; but we have information about both of them now that we did not have before. We have instrumental techniques for making observations of the visible world which raise the estimation of sensible things considerably, and in connection with them we have the scientific method which in a curious and paradoxical way obliterates the distinction which Plato made between the intellectual and the manual.

How are we to know what there is corresponding to rational thought outside the mind? Surely not just anything that we can think about abstractly, for then we should be committed to the existence of fanciful creatures, such as leprechauns, and to falsehoods, such as the equality of 4 and 5, for there is no difficulty in thinking about them. What precisely are the Platonic 'Ideas'?

The whole drive of the experimental method is to separate the leukocytes from the

1 *Republic*, 511 B-C.

leprechauns. Both are Ideas, or if you like, 'universals', a term which is hereby granted a new meaning: extending to all of the members of a particular class throughout the material universe. But as Peirce once observed, nobody ever claimed that all universals are real, only that some are, and he insisted that it is the business of experimental science to show which are (1). For to assert that all universals are real would only weaken the claim for the reality of universals.

The failure of the agonizing attempt to find a material location for the invariants of generality leaves no choice except the positing of an ontological second tier, which the empiricallyminded are temperamentally disposed to reject. But there seems to be no other possible answer. As we noted earlier, the law of gravitation cannot be reduced to the sum of physical bodies attracting each other because of the generality involved in the law. Universals must be the representations of what perhaps may be infinitely many absent objects.

We make direct contact with both ontological orders just as Plato said that we do: with the universals of the intelligible order by means of thought, and with the particulars of the visible world by means of the senses. There are however two improvements which he was in no position to take into the account, and they make all the difference.

The first improvement is that the range of our knowledge of the intelligible order has been greatly extended by means of pure mathematics. The systems of mathematics, algebra, topology and analysis, go far beyond anything the Greeks had envisaged, and this is true even though Plato laid great stress on the importance of mathematics.

The second improvement – and it is a tremendous one – is that *by means of instruments and the new techniques of observation and experiment in the physical sciences we are able to learn from our observations of the visible world about an entirely new set of elements of the intelligible order*. This last proposition is, I suppose, the essence of my argument in this chapter. No advanced technology, no new materials, then no instruments; no instruments, then no knowledge of nature of the kind we have achieved in the last few centuries. The position will require a different approach from the one which had been customary in Plato's day, and it will require also another principle of classification.

The process is not a psychological movement of deriving knowledge from sense experience but rather more like what Whitehead called 'extensive abstraction' (2), which operates on the basis of *the data disclosed* by sense experience and consists in a following of the connections in the external world as they lead from the concrete to the abstract.

---

1 *Collected Papers of Charles S. Peirce*, 5.430.
2 *Process and Reality* (New York 1941, Macmillan), pp. 148, 454-508.

7. The Final Picture

What makes a metaphysics acceptable? Nowadays we ask that it be neither inconsistent with the findings of the experimental sciences nor limited to them. Whitehead demonstrated effectively how the changing cosmology of physics from Newton to Einstein called for changes in metaphysics. It is my contention that the theories of physics and astronomy now prominent require a further revision of the Platonic theory of the Ideas.

If you will look along the time-line of the flux of material existence you will note that all material bodies perish. They come into existence at some date and place, they grow and flourish and then decline and die, and this is as true of stars as it is of people. Each of them differs from all others in some respect, each is unique. Some of their parts, however, such as qualities, relations and forces, and even types of the whole, such as atoms and suns, are similar; so that while they perish, the similarities recur again and again. They somehow enjoy a special status whereby they are saved from the flux, and that special status is what Plato wished to acknowledge by claiming for them a separate domain.

Accordingly I am proposing modifications in Plato's two orders, the intelligible order and the visible world. I have not disturbed the importance attributed by Plato to the intelligible order, but I have raised considerably his estimate of the value of the visible world, placing it first in terms of learning, much as Aristotle might have suggested if he had confined his remarks to what he affirmed.

The visible world is a mixed-up place, for it is possible to discern in material entities and processes in time and space the events which are made possible by elements of the intelligible order — only in this case very much out of order. Chance as much as law sets the conditions for harmonies and conflicts. But actual entities and processes can be not only classified but also controlled and predicted, and that is considerable support for the thesis presented here.

The intelligible order is proving to be far more complicated than anyone had supposed. First of all it is subdivided, and, as we have already noted, there is a hierarchy of logical classes and another hierarchy of empirical laws. The knowledge of each of these subdivisions is obtained by a particular method of inquiry. We learn about the logical classes by means of mathematical calculations, and we learn about the empirical laws by the scientific method. The picture now looks something like this.

THE INTELLIGIBLE ORDER

Objects of pure mathematics
Laws of the experimental sciences
Empirically-verified universals

Table III

The structure of the intelligible order is a standing set of unchanging possibilities. Nothing happens to it in the operations performed by mathematicians. Their formulas merely record how those structures can legitimately be traversed. There is need here for an entirely new theory of traversion, for calculations in mathematics consist in traversing standing relations which remain unaffected by the operations. These are new discoveries affecting idealism: dimensions of essences hitherto unsuspected but hinted at by the high speeds of traversion made possible by computers.

What we are seeking here is of course a highly condensed presentation. It does not display the vast richness of the intelligible order, for instance the profusion of mathematics which is included under the objects of pure mathematics; and it does not give any indication of the distinction between generality and inclusiveness: the greater inclusiveness of mathematical laws, for example, over the inclusiveness of empirical laws with which they share an equal generality. And of course — and perhaps worst of al — it does not give any hint of the richness of the differences inherent in the mixed-upedness of the material world with its harmonies and conflicts, its regularities and disorders. I hope to repair these deficiencies in Table V.

First, then, with regard to the natural order.

The 'objects of pure mathematics' need no help. Plato had already located these among the Ideas.

The 'laws of the experimental sciences' presents a different case. The laws discovered by means of observation and experiment are invariants; they are fixed relations with an empirical content. They do belong therefore to the intelligible order, and with them it could almost be said that Plato's formulations would hold without any additions.

Not all empirically-verified universals are included in the above. They are everywhere typical, however, and there is no established holotype. Plato's attempt to set up good-ness or beauty as holotypes only led to metaphysical confusion. It is just possible of course that eventually they will be included among the laws of the experimental sciences.

Next with regard to the material world, which displays a greater variety than even Aristotle envisaged.

The 'contents of the integrative levels in disorder' would presuppose a knowledge of the references of the integrative levels. These are: the physical, the chemical, the biological and the cultural. They have their own order though it is not displayed. What is displayed are the elements of the integrative levels *in disorder*.

The 'properties of matter' need little more explanation than the reminder that matter is interconvertible with energy. The connection between energy and the passive qualities is perhaps less well known and would require a further explanation.

The 'states of matter' — one of them, at least — is perhaps less familiar. The plasma state, which, as we have noted, is the condition of excited ions, or electrified gas, constitutes the greatest quantity of all the matter in the universe. We did not until recently possess the instruments which would enable us to discover that this was the case, but it does mean that now all theories of materialism will have to be radically revised.

CHAPTER IV

UNIVERSALS: THE QUALITIES

1. The Failure to Recognize Force as a Quality

In chapter II, section 5, above, there was a brief discussion of qualities, and I hope that I was able to establish there that qualities are kinds of universals. In this chapter I will have to focus disproportionately on the much neglected fact that force, too, is a quality. This aspect has been neglected; in the account of qualities throughout the history of western philosophy qualities are usually described as passive affairs.

It is simply astonishing that, with the possible exception of Leibniz, the traditional European philosophers omitted all consideration of force as a quality despite its prevalence. The omission seriously damages whatever explanatory value the discussion of the other qualities may have had. Leibniz, it is true, did have a large place in his system for force. His 'monads' were units of force, but oddly enough they were 'windowless' and without interaction, the appearance of interaction being accounted for by a 'pre-established harmony'. In short his units of force exerted no force.

The qualities treated by the empirical philosophers of the seventeenth century were assumed to be inert. The world as reported by them was substantially the one described by the rationalists. Whether knowledge was a product of sense experience or the result of reasoning, the conclusions were similar, but they were never characterized for what they were' the account by a passive individual of his experience in the midst of a passive environment.

The elements of the world so discovered were given in subjective terms called perceptions and thoughts, and were those of an individual who seldom moved or was moved. His perceptions came from particular senses, chiefly sight and hearing but also taste and smell. None involved activities of any sort, so that a philosopher was able to sit perfectly still and yet make all the necessary observations. Locke, Berkeley and Hume were mostly concerned with passive sense experience, and much the same could be said for Descartes, Spinoza and Leibniz, who put together a rational account of the individual and his world. Movement was suggested by Berkeley, but his proposal was negligible, and was never extended as it might have been.

The sense perceptions of the empirical philosophers gave rise to the knowledge of qualities, and the thoughts of the corresponding set of rationalists gave rise to the knowledge of relations. Qualities and relations, these were the units dealt with most, and it was not noticed that they were as inert as the subjects to whose senses and mind they were supposed to be attached.

Evidently for all of these philosophers the ongoing world of activity and conflict was only an appearance whose reality was made up of passive units. the mind of the knower was supposed to contribute a large portion of what was known, so that in a subjectively-oriented perspective it was considered theoretically possible to inspect the validity of the contents of knowledge by an examination of the working of the mind.

The shortcomings of idealism were never more in evidence. Ordinary sense experience unaided by instruments not only gave a limited version of what the material world is like but was acutally misleading. The world was expected to conform to the ideas which were held about it, and nobody bothered to check out the conformity except the scientists, whose conclusions were not admitted to have any bearing on philosophy.

The scientists had as a matter of fact followed a different path; for whereas the empirical philosophers were occupied with the subjective end of sense experience and so sought in the mind of the subject a clue to the nature of reality, the scientists were concerned with the objective end and thus landed not in the experience itself but in what it disclosed to them about the material world.

How did it happen that the philosophical empiricists of the eighteenth century overlooked the one important sense organ which held the clue to such a large area of existence? I allude of course to the skin. The understanding of the quality of force might have come first from the generalization of pressures on the skin as the one sense which is in contact with that element of the world which offers resistance. It might have made more difficult the subjective interpretation, offered by Locke for it might have led him to see that the defence of the theory of sense objects as products of sense organs had been rendered incomplete. The resistance to pressure is not likely to be offered by a substance that is not there and could not long have remained a 'something, we know not what'. In forceful activity, bodies affect each other as well as the knower and do not depend upont being 'permanent possibilities of sensation' as Mill thought.

Then again, why was it that no one seemed to have noticed that the coordination of the senses – including the sense of touch – produces a conception, not otherwise obtainable, of a world of stereoscopic objects possessed of diverse properties and interacting with each other in bland disregard of the subject? For a material object to be successively seen, touched, smelled, tasted, and collected into a conception of the whole, either it must be moved or the subject must move in relation to it, and in this way some knowledge of it might have become available to theorists whose reliance upon single senses has misled them so badly.

If it was clear all the time that the world was not like the one described by the philosophers, nobody raised any objections. In the nineteenth century it is true some did, but in a certain connection these later thinkers continued to miss the point. Schopenhauer, still hopelessly enmeshed in the Kantian orbit, talked about the 'will', by which he meant to describe and to generalize the kind of force which exists at the physical level, but that did not get him out of the charmed subjective circle.

Nietzsche was perhaps the first European philosopher of any moment to acknowledge the importance of force. Unlike Schopenhauer, however, he saw it exclusively in socio-cultural terms. What a different turn philosophy might have taken had either Schopenhauer or Nietzsche recognized the importance of artifacts. But both men were victims of that subjective orientation which plagued philosophers and did not allow them to come to grips with the objective and independent nature of the material world in which man-made objects play such an integral part. Culture for them was purely a personal affair and they saw it as an element of society into which it had been projected by individuals.

The qualities as the philosophers have traditionally understood them are the familiar ones corresponding to the senses, the senses of taste, smell, hearing, sight, the four so-called skin senses: touch, warmth, cold and pain, and the organic sensations, such as hunger and thirst. Curiously, they knew about but did not deal with the muscle sense, which issues from nerve endings deep in the muscle fibers and records the weight of objects, their hardness or softness, and, most important, their *resistance*.

So preoccupied were the philosophers with the senses themselves that they forgot to notice what those senses report, understandably when we remember that the philosophers tended toward epistemological idealism. The muscle sense has not received even yet sufficient recognition as the one which corresponds to the existence of forces. Force is a sense quality, resulting from the combination of skin touch and muscle tension. It records the existance of external events.

That force is not only a quality but a primary quality is witnessed by the fact that we all have material bodies and exist in a world of other material bodies, with some of which we make contact. An individual who was deaf, dumb and blind could still feel physical encounters with other bodies, could still have the sense of pressure, could still experience resistance.

Force, then, provides the entering wedge to the understanding of much in the nature of man and the world he inhabits. Force as a physical effect has not been neglected, its nature as a quality has.

2. The Properties of Qualities

Before I can argue that force meets all the requirements of a quality, I must first summarize what is known about quality. Let us look next then at quality with all of its attributes. Once its nature is established, we can more easily examine the properties of force as a quality.

Elsewhere I defined quality as that which is ultimately simple, on the assumption that qualities as such are indecomposable (1). I still hold that to be true without denying that qualities do have external aspects. For qualities possess properties, they have duration, spread (or extension), strength, dependence, insistence, and possibility. These of course are not analytic elements but features the qualities have as wholes.

A quality is neither a universal nor a particular. It is not a universal though universals are classes and quality is not a class nor a member of a class; but neither is it a particular, for particulars are unique with respect to both date and place. Yet it has similarities to both, and these may have caused the confusion. Like universals, qualities recur, and it is notable that when they do they are always the same, which makes them seem to be eternal; like particulars, they are characterized by opposition, historicity and infinity, that is to say, like concrete individuals, qualities are stubborn. They resist all efforts to make them other than they are; they inherit their own past occasions of endurance; and they disclose the irrefrangable fact of having existed whenever they did (2). Thus qualities constitute a third kind of entity, apart from universal and particular yet partaking of both.

Qualities have their own peculiarities. For instance, every quality is independent of all others (since no two qualities are exactly alike); with all qualities the part is equal to the whole; any small amount of a quality is just as qualitative as any large amount, as for instance a small red patch which is as 'red' as a large one; qualities cannot be measured by quantity, for any two amounts of a quality are the same qualitatively (which makes qualities into continua; qualities as such are not structured, only limited by structures, with qualities at every structural level).

Generically speaking, qualities too are iterations. This is essentially a new idea, and so I had better say a further word about it. Whenever and wherever a quality appears, it is always the same. Different appearances of a quality are *identical* rather than *similar*. The recurrence of a quality is an identity, and identities are not relations (3).

Finally, there are the qualities which emerge at every integrative level of organization, discrete qualities which lie along a continuum marked by organizational breaks.

1 *Foundations of Empiricism* (The Hague 1962, Martinus Nijhoff), p. 78.
2 *Op. cit.*, p. 61.
3 See above, chapter V, Section 2.

What we have ordinarily called qualities, those suggested by sense perception: colors, tastes, sounds, are in fact qualities at the physical level, but there exist also chemical qualities, such as valence, biological qualities, such as life, psychological qualities, such as consciousness, and cultural qualities, such as values.

We will look at this structuring of qualities in the next chapter. Here it will be necessary only to point out that qualities at the same level are ordered: some are compatible, others not. Some colors, for example, blend very well, while others clash. But colors are always attached to material things and never occur alone, and they always belong to the present.

3. The Quality of Force

My next task is to show that the quality of force answers to the above description of the properties of qualities in general.

The quality of force has the same properties as the others, it endures in space and time, it extends over areas of materials and events where it is concentrated and exerts pressure. Force is neither universal nor particular, although sharing the properties of both; that is to say, it is universal in that it can always occur, and particular in that when it does occur it is always at some specific place and date; but it is not a specific thing like a material, and not the member of a class.

Every force is independent of every other, since no two are exactly alike; a small force has as much the quality of force as a large one, but its quality cannot be measured even though its amount can. As qualities, forces cannot be structured, they do not possess parts. Like other qualities, forces are iterations. Forces at the same level may be opposed, they are countervailing. They are always in the present and always attached to material things.

We have noted how everything that has been said about qualities in general applies equally to the quality of force, but force as a quality has properties not possessed by the other qualities. These properties make it necessary to discuss the quality of force as a separate and distinct quality.

The stuff of the world, its substance, consists in qualities and energy bundles or forces. Qualities are always described as though they were passive, but forces, as we have noted, are active. Qualities and forces are the qualitive correlates of matter and energy. Quality is the passive phase, force the active phase; both belong to the same category of qualities, to which unfortunately the passive phase has given its name.

Forces, then, may be regarded as active qualities. They are the entering wedge to an understanding of nature. All material encounters are felt as forces. Some lie within the human, or mesocosmic, range, but most do not. This is a matter both of the intensity of pressures and of the size of the units on which the forces are exerted. Exploding stars in the macrocosm and the short-range forces within the nucleus of the atom (the microcosm) lie well outside the human capacity to feel. But that surely does not mean that they are not qualitative. When we say that a bolt of lightning cracks a stone or that a bulldozer digs a ditch we are talking about the exercise of force. There is quality in both instances even though it may not be felt as such. Ideas are forces which compel the efforts to bring events into conformity. It can also happen that the force of ideas is felt without an awareness of the ideas, as when a feeling of conviction exists independently.

Only forces in a narrow range are felt as qualities. Indeed the impact of a small force may be felt, while that of a much larger one may not be. This is a matter of the intensity of pressures. A man could feel it if someone punched him but he does not feel the impact of cosmic rays.

We recognize that most of the forces in nature are beyond human control. Every violent thunderstorm releases energies of enormous dimensions, and the same can be said for earthquakes, tidal waves, and other such phenomena. Engineering may be described as the design and application of large-scale artifacts intended to bring forces to a focus. It marks the attempts to operate with some of the forces of nature in order to place them at the disposal of human enterprises. But of course there are always forces at work many times larger than any which lie within prospective human control, such as those of galactic dimensions. It is useful to recognize that these must be qualities, too.

Since most of the astronomical universe is in a state of commotion, force has to be reckoned with as a principal ingredient, and where there are forces there must be also the quality of force. In previous cosmologies this has not been counted, but it should be included in any reasonable reckoning. Matter in motion in space and time is matter propelled by force, and given the plasma state, the importance of force can hardly be over-estimated. Matter in any of its states is always exerting force or being worked on by force; and if we take its convertibility with energy as a fact, then matter may be said to consist in force, and remain potential until actualized in the exercise of energy. Force is therefore the quality which chiefly prevails.

4. The Background of Violence

We are rapidly learning that the traditional picture of the sky is a misleading one. Due to a bewildering accumulation of new discoveries in astrophysics, it now appears that the quiet universe believed in by our ancestors, with its night of silent stars in their courses, no longer fits the facts. Instead the scene is one of great violence, containing forces so tremendous that we have as yet little understanding of them. Stars are hardly serene affairs; they are known to collapse, they explode, they pulse, they burst, they spin and eventually they become extinct. They are probably also very noisy. Theoretically it can be hazarded that sound waves must be generated by the turbulent motion of the gasses in their atmosphere; we know for instance of the existence of high velocity stellar winds and of shocks of great velocity that occur in the interstellar medium (1); but we are unable to hear them because sound waves, unlike light waves and radio waves, can travel only through a medium, and so do not cross empty space.

There might have been an earlier hint of this in the nature of the sun, which is the nearest star and upon which the earth and its system of life support entirely depend. For the sun is anything but an inert body. Astronomers see a massive ball of fiery gases, and have noted chiefly some of the characteristics of a disturbed outer region, with sunspots and solar prominences flaring up many thousands of miles above the surface.

There are many places in the universe where there is matter without life, but none where there is life without matter. Man himself exists on the earth in a relatively peaceful corner of the universe of exploding stars and colliding galaxies; but even there it is not always so quiet if you count earthquakes, volcanoes, hurricanes, tornadoes, and extremes of heat and cold.

Current theories of the origin of the planets place them in the neighborhood of the young sun where they were formed from eddying dust and gas. Matter is transformed into energy at the very high exchange rate calculated by Einstein in accordance with the law he discovered, $E=mc^2$, where E is energy, m is mass, and c the velocity of light. In short, enormous amounts of energy are released at the cost of very little mass.

All matter in the universe is in a continual state of activity, and activity means that forces are being exerted. The physical laws are descriptions of how forces act. This would not be possible if there were no observed regularities. To say that we recognize a physical law is to say that we know what the effect of a particular kind of force will be under given circumstances, how the relevant phenomena will behave.

It has been calculated with more or less accuracy that in a few billion years the sun will become a red giant, expanding to engulf the planets and to occupy the entire region of what is now the solar system. It will then shrink to the size of a white dwarf and become a black hole.

The same fate of the expanding sun, which meanwhile burns at a furious rate, awaits the billions of other stars in our galaxy and in the billions of other galaxies scattered throughout the universe. Astronomers agree that the Crab Nebula shows today the debris left by a star that was seen to explode in AD 1054, and which is still a source of considerable radio noise. The big explosions of stars, marked by what are now called supernovae, have been observed three times in the past thousand years, two in addition to the Crab Nebula, one in 1572 and another in 1604, often enough to tell us that it is a common occurrence. Stars do explode, we have sufficient evidence of that

1 *Ann. Rev. Astron.*, 1979, p. 213-240.

now, and consequently any planets in their neighborhood would be adversely affected. Violence in the cosmos is the rule, not the exception.

Violence in the life-cycle of stars can be matched by the violence observed in galaxies. Just as exploding stars have been found in abundance, so have examples of exploding galaxies. Many of the galaxies are in a state of extreme disorder. Exploding galaxies are not unknown, there is for instance $M82$, a galaxy which appears to be shooting large streamers out from both sides, with a center which is a source of strong radio noise.

Variations observed in the brightness of galaxies point to events of great violence involving millions of stars. One good example is the class of galaxies known as seyfert galaxies after Carl Seyfert, the astronomer who discovered them. The seyfert galaxies are suspected of having some connection with quasars though they are not quite as bright. Quasars, with centers of power hundreds of thousands of times greater than the sun, remain unexplained. The seyfert galaxies — all twelve of them — have been found lately to emit more heat than light, additional evidence of violence.

Violent stars, violent galaxies, and finally, a violent universe. The universe itself may be said to be expanding, or exploding; that is to say, all the spaces between all of the galaxies are increasing: empty space interspersed with violent matter. The ongoing theory, that the universe as we know it began 13 billion years ago with a 'big bang' in which all matter was involved and which resulted in the current expansion that sees all material bodies rushing away from each other, has been reinforced by the discovery of a weak but persistent microwave radiation, radio waves a few centimeters in wavelength coming from all parts of the sky at all hours of the day and night, called K black-body radiation because it is three degrees above absolute zero.

An alternative to the Big Bang theory is the theory of the oscillating universe, that the universe pulsates in multi-13 billion year cycles. All of the recent phenomena that have been discovered, such as the quasars, the pulsars and the black holes, whatever else they do, give evidence of extreme violence, and of great forces at work.

5. The Quality as Universal

Perhaps the most commonly encountered universal is the quality of force. It is the essence of all activity, which usually involves overcoming resistance, at least the one supplied by the force of gravity. We deal in all of our activity with exemplars of qualities, all qualities, not merely the quality of force. We are guided in many of our actions by the sense qualities, which are passive, but the amount of violence in human affairs should tell us something, wars having been that prevalent.

All encounters with force by human individuals make it seem intensely particular, yet it is only and always an iteration of an universal. From the merest breath of wind to the atomic bomb, we are encountering iterations of the universal of force.

I have emphasized in this chapter the quality of force as a universal because it is the most neglected, but the more passive qualities also exert force though in a milder and often more subtle form. We are guided in many of our thoughts, feelings and actions by the effects on us of the sense qualities. The smell of perfume, the smooth touch of silk on the skin, the taste of agreeable foods, all of these are qualities and as such universals. The universal nature of even the sense qualities has been much neglected; we never think of them as exemplars or as iterations, only as things in themselves and hence particulars. Of course they do have the aspect of particulars, and that is what captures attention; but they are also and essentially universals.

The similarity, indeed the identity, of an instance of a sense quality with others which are like it is seldom recognized because the quality is what it is and not another thing. Passive sense quality or active force, the quality as universal has an important property which it shares with all other universals, such as the forms. They do not perish. Qualities and relations may be the only things in the cosmos that do not come to an end. So long as there is anything there will be qualities and relations. They are the only truly independents. When they exist they are attached to material bodies which occupy space and time; yet if they do not continue forever any more than individual material bodies do, they have the virtue that they recur whereas individual material bodies do not.

CHAPTER V

THE UNIVERSE

1. The Atypical Earth Conditions

The rapid development of scientific cosmology in the last few decades is the product of experimental physics and astronomy. It suggests the need for a broader and more inclusive consideration in philosophy. The aim of philosophical cosmology is to take into account many of the suggestions made by the scientific cosmology and extend beyond it without becoming inconsistent with it. Physicists, astronomers, chemists and biologists have furnished a groundwork of observations resulting in theories and facts, but the philosophical cosmologist must find a broader conception in which these can be included as special cases.

As we noted at the end of the last chapter, the study of cosmology has been very much hampered in the past by the fact that the conditions found on earth by means of the unaided senses were regarded as typical. Previous cosmologies were tailored to terrestrial conditions, informed only by naive experience, all local, many misleading as cosmic indicators. Now, instead of interpreting events in the cosmos in the light of what we know of events on the earth, we shall have to learn how to interpret events on the earth in the light of what we know of events in the cosmos. (I am not of course suggesting that events in distant space can influence the course of human lives here, only that events in distant space being more numerous should be regarded as more typical.) Although the natural laws called out by conditions here must be everywhere the same, conditions elsewhere may call out natural laws not called out by conditions here. Consider for example the kinds of regularities which must exist in such celestial objects as galactic nuclei, quasars, radio galaxies, pulsars and supernovae. The prevalence of violent events in the cosmos, events which would render all life impossible if they occurred on or near the earth, attest to that fact. The probable existence of life on some planets in other galaxies simply means that this sort of singularity too may be common.

Let us see, then, how the new scientific cosmology affects the older studies of epistemology and ontology.

2. Epistemological Implications

The acquisition of knowledge begins with the simplest of the sense perceptions of the subject. These are connected with external material objects by means of qualities and relations. At the end of chapter III, I pointed out that the reality of universals is forced on the observer by the fact of his recognition that the material object as a member of a class of such objects is a prerequisite of his perceiving at all: universals are perceived in things as their classes and relations. Then also in chapter VIII, Section 3, I will show that the recognition of an object involves the double perception of the object and its class, a class which, incidentally, includes an indefinitely large number of absent objects. There is, in short, direct contact with universals.

This conception is by itself difficult enough, but unfortunately there is more. It has been an assumption of naive realism that everything must rest on *unaided* sense experience, and that assumption is still being made. It has had a long history and served its turn well because, empirically at least, there was no alternative. The assumption, however, no longer prevails, and so we will have to take a somewhat different view and prepare to rearrange our priorities. For the new knowledge is of objects too small or too large to be reached by unaided experience.

To obtain such knowledge the scientist employs instruments with more acute perceptions than his own, data which come to him in the form of pointer readings, tapes or photographic plates. One recent variety, remote sensing, as it is sometimes called (1), enables the scientist to collect information about an object though lacking physical contact with it by using sophisticated instruments that can detect electromagnetic energy in one of its many forms, such as visible light (2), infrared radiation or radio waves. The instruments used in other observations are in a way extensions of sense organs. I am thinking of such examples as photographic emulsions, thermo-couples, photon tubes, oscilloscopes, transistors, cloud chambers and scintillation counters.

As a result of this technology science presents epistemology with a grave new problem. The ordinary world which confronts us in our efforts to acquire reliable know-

1 Floyd F. Sabins, *Remote Sensing* (San Francisco 1978, W.H. Freeman & Co.).
2 For what can be seen just by the use of visible light by cameras borne aloft by spacecraft, See Oran W. Nicks (ed.), *This Island Earth* (NASA SP-250, Washington, D.C. 1970). Remote sensing is well described in this book, pp. 164-166.

ledge of it — let us call it the mesocosm — is no longer the simple one known to previous epistemologists. For what is available to our sense experience has been found to be bordered on both sides by contiguous extensions. It is not an accident that 'the mass of a man is the geometric mean of the mass of a planet and the mass of a proton' (1). The microscope has disclosed to us the existence of a world of very small objects, and the telescope correspondingly a world of very large ones. Knowledge of the objects has been expanded still further by employing sophisticated versions of these instruments: the scanning electron microscope, for instance, and the X-ray telescope. We may call the world of the very small the microcosm, and the world of the very large the macrocosm.

What we need is a set of concepts which together could provide 'a close linkage between the largest-scale structure of the universe and the world of everyday experience' (2). That linkage is provided by the recent extensions which have stretched 'common experience', the old name for the disclosures of material objects by the unaided senses.

The simplicity of the old theory of matter was misleading. That the universe is far more complicated than had been supposed is now made evident by the enormous progress that has been made in the analysis of matter. The resulting complexities were suggested in Table I on page 26, above.

THE THREE-SEGMENTED MATERIAL UNIVERSE

THE MICROCOSMIC	THE MESOCOSMIC	THE MACROCOSMIC
microscopic objects; atoms, electrons, etc.	objects available to ordinary experience; tables, trees, etc.	telescopic objects; suns, planets, galaxies

Table III

Cosmology is an extension of the theory of matter. The new picture of the external world shows it to consist in a three-segmented universe, each segment having its special properties. Prolonged and detailed investigations by various groups of scientists, each with its own specially designed technological equipment, disclose a three-seg-

1 B.J. Carr and M.J. Rees, *Nature*, 278, 605 (1979).
2 Fred Hoyle, *From Stonehenge to Modern Astronomy* (San Francisco 1972, W.H. Freeman), p. 75.

mented world divided into what we agreed to call the mesocosmic, the microcosmic and the macrocosmic segments, respectively. The mesocosmic segment is the world of our ordinary experience as disclosed to our unaided senses. The microcosmic segment, with its various sublevels, is the one studied by nuclear physicists, physical chemists and molecular biologists. And the macrocosmic segment is the one observed by astronomers and astrophysicists. Each segment has its own mechanics: classical mechanics for the mesocosmic segment, quantum mechanics for the microcosmic, and relativistic mechanics for the macrocosmic. Correspondingly, each segment has its own measurement of time, clock time for the classical, atomic time for the microcosmic, and astronomical time for the macrocosmic (1).

The three-segmented universe is an entirely new conception and it presents many problems for philosophy to solve. Some of the solutions can be conjectured; indeed the whole of cosmology presents a conjecture, one which, however, is contained in a theory suggested by fact. There are problems presented in epistemology which cannot be solved at the present time even by conjecture.

The crucial fact is that the contiguous areas, the microcosmic world on one side of the ordinary world and the macrocosmic world on the other side, are without a break of any sort. This is made evident by the continuity of the three-segmented world, each segment containing its own integrative levels, which all together constitute a unity.

One of the key conceptions of the new cosmology, then, is the continuity which prevails between the microcosm and the ordinary familiar world (or mesocosm) on the one side, and between the mesocosm and the macrocosm on the other. There is no philosophical background whatsoever for this conception simply because the existence of the two adjacent domains were unknown until comparatively recent times. Many important implications remain to be explored. The material universe is considerably more extensive than had formerly been thought, and the consequences of this to metaphysics as well as to epistemology will have to be investigated.

Science assumes a basic realism for the perceptions of objects in the ordinary world. For objects in the microcosmic and the macrocosmic segments, then, the perceptions will be obtained from configurations on recording devices. Henceforth we will be compelled to rely upon the superior perceptions of instruments by means of which material objects smaller or larger than those ordinarily available to the unaided senses are rendered perceptible. It is crucial to the argument, however, that these must be interpreted in the ordinary way. We have to remember that what is seen on the plates or counted by the pointer readings is also out there in the world. The instruments function as surrogate senses. Images and quantitative constructions for the first time become a certifiable part of knowledge and an index to the equal reality of the whole three-segmented material universe.

1 On the last, see P. Kartaschoff, *Frequency and Time* (New York 1968, Academic Press).

The epistemological assumptions underlying the scientific method of realism will mean that the abstract laws as well as the material facts are understood to be imbedded in the regularities of phenomena and as such assumed to be independent of the knowledge of them. The scientific investigator is occupied with observing the experiment he is conducting, not his own activities. Of course he does have reactions which are subjective, but they are not what he is after. He counts on them only for what they can disclose to him. He observes the world; if he were to observe also his own reactions except in the case where these are the sole targets of his investigations this would involve a double rather than a single subjectivity, which is unwarranted by the aim he has chosen to pursue. He employs his own reactions to tell him about the world, not to tell him about himself.

Let us suppose that he projects a beam of light onto a scintillation counter. What he observes are the pointer readings. He certainly does not observe himself observing them. The presumption is that the two material objects whose reaction he observes are independent of his observations, for if they were not why would he go to such experimental lengths to obtain them? He has simpler ways than that of finding out about himself. Experiments are designed to yield a certain kind of answer and conducted to discover just what that answer is.

At the present time it is impossible to develop a complete conception of the objects of the microcosmic segment of the world in all their full concreteness. We have at our disposal only visual impressions, that is all the photographic plates can tell us. We know some of the relations and forms, some of the structures of objects, in that world, but we know nothing of what its other properties are like.

Suffice at this juncture to say that the realism which the theory of knowledge presents raises an important point. There must be a continuity of objects in the ordinary world (i.e. the mesocosmic segment) with objects in the microcosm on the one side and objects in the macrocosm on the other. If there is a three-segmented universe composed of matter and energy existing objectively to and independent of all our perceptions of it, then the conditions described by realism must extend to objects available to us only be means of instruments.

Can we justifiably argue by analogy that if objects in the ordinary or mesocosm possess colors and odors, make sounds, and offer resistance to pressures, then objects in the microcosm and the macrocosm must possess them also and in much the same way? It would seem that we can. Quarks and gluons in the microcosm, quasars and pulsars in the macrocosm, must be like objects on the earth's surface with respect to the sense qualities. Do they have properties corresponding to what we should describe as color, smell or taste? Analogy allows us to reach no different conclusion.

Common experience, despite the fallible senses, has provided a conception of the external world which accords ill with that picture of it the philosophers had constructed, a picture such as Locke's in which an impoverished nature devoid of all senses qualities could boast only of the physical properties. Science has radically altered this conception. It is, paradoxically, the physicists who have led us to understand that the real world is an immensely rich affair, a dense universe of complex materials moving in space and time and containing much more than our senses have disclosed.

The evidence comes from those two adjacent segments, whose properties, added to those of common experience, reveal a universe having a degree of intensity as well as dimensions of extensity we are only just beginning to probe. The degree of intensity is shown by the way the world is built up layer by layer of accretions of levels of complexity, with structures germane to every level. For the dimensions of extensity we shall have to turn for evidence to recent advances in astronomy.

At the present time our only knowledge of the cosmos comes from the feeble perspective taken from the earth. It comes in two forms: (a) what we can apprehend from the earth in distant space, and (b) what we can regard as typical on the earth.

(a) What we can apprehend from the earth is given in astronomy, employing telescopes both optical and radio, as well as instruments for other wave-band lengths, together with photosensitive plates of all sorts used with different filters and other attachments. In recent decades more instruments have been added to the arsenal, and each new instrument provides new data taken from a new perspective; for instance the radio telescope can make possible four measurements of any radio source: flux, density, position, brightness distribution and polarization (1).

The scientific instruments employed to probe into those segments of the cosmos which are unavailable to the unaided senses should not be regarded as we regard prosthetic adjuncts of the human body, such as artificial hands and legs; yet there is something to the analogy, for, without giving it a subjective reading, telescopes and microscopes do extend the senses into regions not otherwise accessible to them. At the same time, what is disclosed by these instruments is as independent of them as it is of the observers without them: a world of matter and energy in motion in space and time which is no wise depends upon human knowing.

What is called 'scientific knowledge', and looked upon in some quarters as the exclusive property of the scientists, arcane to the uninstructed, is in fact the knowledge of a cosmos nobody owns. What is scientifically discovered is not scientific by nature, and what we learn from the science of astronomy for instance is not about astronomy but about the cosmos. The discoveries of astronomy do not belong to the science of astronomy. Astronomy is in this sense only a window on the world, and what its practitioners have discovered are world conditions as these are disclosed by instruments designed to detect electromagnetic waves of various lengths. What emerges is

---

1   *The Structure of Galaxies*: Proceedings of the Thirteenth Conference Physics at the University of Brussels, September 1964 (London 1965, Interscience Publishers), p. 108.

the picture of a universe of great complexity, containing a variety of objects and energies of enormous dimensions.

The only information that comes to us from remote space is in the form of radiation of one sort or another, yet it has been estimated that there is in the universe 'something like 100,000 times as much matter as radiation' (1). This means that for the greater preponderance of matter we have no information, and consequently that there is an imbalance in the information we do have which has not been properly recognized: it comes from too small a sample.

(b) What we regard as typical on the earth introduces a serious qualitification which must now be mentioned. We usually take our readings from some humano-centric perspective: how items on the earth appear to us, what they mean to us, or what they can do to or for us. But perhaps there are other lessons to be learned if we view the same items as cosmic indicators by considering what part they play in the universe considered as a whole. Among these, force is the most prominent, that is to say, force considered as a quality (2), together with other qualities, relations and structures which enable it to be brought to bear.

3. Ontological Considerations

Epistemology will be kept in the picture as a partial explanation as long as the observations are not complete, and this of course means indefinitely. Meanwhile, however, there remains in the background and necessary for such operations the outlines of a probative ontology, a system which is assumed by the proposal of hypotheses and by the methods of testing appropriate to their investigation. No one would seek the knowledge of what he did not already assume to exist.

The outlines of such an ontology are not far to seek. We have already noted, thanks to modern physics and astronomy, that matter is capable of supporting many levels of organization, from the very small units (quarks, protons, neutrons, electrons, atoms, molecules) already referred to as the microcosm, to the very large (planets, suns, hydrogen clouds, galaxies) of the macrocosm, with many levels between these extremes (plants, animals, human societies), all belonging roughly to the mesocosm.

The important point is that no material organization at any one of these levels can claim ontological superiority over the others. The cosmos is so large and the local conditions so repetitious that the organic life which prevails on the earth is probably repeated elsewhere many billions of times. The thesis that the purpose of nature is to produce man certainly does sound like a piece of special pleading. What we are faced with is a changing world of material particulars in which there can be discerned and

1 F.P. Dickson, *The Bowl of Night* (Cambridge, Mass. 1968. The M.I.T. Press), p. 186.
2 See above chapter IV.

abstracted the equations whose reference is to a recurrent world of forms, qualities and relations, together making up a two-tiered universe of actualities and possibilities engaged as staging areas in a repeating and continuing process.

Existence is comparable to a very large game in which all of the rules but none of the moves are determined *a priori*. It is played at a number of interacting levels and with many different sorts of counters. That these are ordered in a complex framework is made more and more evident when the permanent elements are seen to be arranged in a stratified series of structures from the simpler (though still enormously complex) to the more complex still.

It might be helpful if we could discern in the material particulars of all three segments of the universe some of the conditions that such structures have in common regardless of their size or degree of complexity. They have in common that they share continuity, plenitude and gradation, that they are located in space-time; that they are composed of matter or energy and can participate in things and events; that they suffer privation, discontinuity and inequality; that they are subject both to cause and to chance; that they are made up of wholes and parts, with attractive and repulsive forces acting between the wholes and between the parts; that there are periods of stability and equilibrium, so that every whole and every part has a career however short or long, and goes through a cycle of fixed frames of destiny.

There is a sense, then, in which ontologies are not pictures but glasses of reality, intended not to be looked at but to be seen through. The proposal of an ontology is that it serve as an aid to theory and observation, to assist in placing the conclusions arrived at in this way in a larger context in which they become more meaningful. The design of an ontology is a professional task, one for which it may be unnecessary to go outside the technical terms and their manipulations; but this is not true of the result, for we are dealing in this chapter not with a complete ontology but only with that half of it which is named cosmology, the half which consists in the three-segmented existential universe. We want to learn about its structure of energy-levels: what they contain as well as how they are connected, using the knowledge collected in all of the experimental sciences to date, with the further assumption that additional data would serve only to round out the pattern without compelling its exchange for another.

We have been looking at matter in the ground state, that is to say, matter at the lowest levels of organization, which are those treated in physics. There are other levels of organization which are constructed over the ground state and so depend upon it while rising above it, in a series determined by complexity and, more importantly for our purposes, by emerging qualities. These levels, already referred to, are: the chemical, the biological, the psychological and the cultural.

The continuity which links the three segments of the world together suggests that a radical difference in kind is not likely. Chemical elements and compounds have been detected in the interstellar medium, for instance, such substances as oxygen-hydrogen molecules, carbon monoxide, water, ammonia and even formaldehyde. And there are other and stronger reasons for believing in such continuity, for instance the fact that the very small-scale objects furnish the building blocks which enable the organizations of middle size and of very large scale to be constructed. The bricks in a building do not cease to be bricks when they become parts of containing walls. We shall see that the atoms of physics when many are involved make up the molecules of chemistry, and when the molecules of chemistry are arranged similarly they make up crystals and organic cells, and cells make up organs and finally organisms. Each of these structures exists at a certain level of organization, and because each is a sensitivity-reactivity system we are entitled to call them energy-levels.

The sum total of all sensitivity-reactivity systems at various energy-levels which exist together and interact as a unit is called 'the cosmos'.

So much for the level of matter, which has a richness furnished to it by the wealth of inconsistent subsystems. The systematic consistencies, which are to be found among the larger of the elements of the recurrent order of similarities, are inadequate to describe the world whose greater completeness is made possible by the paradoxical inclusion of conflict. In short, more than one consistent system is required to complete the description of existence, because of the ingredient of opposition. When we try to represent opposition abstractly, we run into severe difficulties, for we encounter the effort to describe contradiction as the abstract coefficient of actual opposition, and so we are left with the problem of embracing inconsistency within a consistent system.

4. The Two-Tiered System of the Universe

In the end we are led to suppose that each tier of our two-tiered universe has its own quite different characteristics: the universe of the transient sequence of differences (material entities and events), from which we shall have to filter out the recurrent order of similarities (laws, qualities, forms and forces). The transient sequence of differences is the familiar visible material world of actions and reactions, of forces and qualities, of movements and conflicts, the world of common experience. The recurrent order of similarities, or the intelligible order, is the structure of universals and universal laws, as discovered by mathematics and the experimental sciences respectively. We know the two orders better perhaps as the visible world and the intelligible order, though the former is not all 'visible' and the intelligibility of the latter often difficult to discern.

The visible world, discussed in chapter II, and the intelligible order, discussed in chapter III, are here brought together and related. The kinds of forms can be recognized: the intelligible order is made up of those elements that are to be found throughout all of space without any restriction, i.e. classes of classes in various combinations; while the visible world contains also those that are under specific restrictions, i.e. classes of material objects.

It would seem that there is no ontological empty space between the two sequences though there is a separation, for it does not have to provide a no-man's land where supposed integuments would strive mightily to bring the two into some sort of concord. The most logical of abstractions are always available from their position, while the most material of concretions disclose uniformities; and in this way the two shade off into one another.

There is another way to look at the same phenomena. Due to the presence of energy, all matter is in a constant state of change. All change involves the exchange of forms (or universals). Therefore energy is equivalent to matter exchanging its forms. The relations between the two domains, then, are initiated by material events.

I do not need to have it pointed out to me that this interpretation is controversial. But I would argue that there are things to be said about the visible world and the intelligible order when considered together that can be said without violating the canons set for experimental and logical evidence. The procedure whereby the elements of the intelligible order occur in the visible world show that world and its order intimately related. Together they make up the system of the universe.

There is evidence that such a system was not unknown to Plato. Those who sum up his metaphysics by an account of his theory of the Ideas as set forth in the *Republic*, often fail to provide a proper place in it for his theory of the material universe, as given in the first half of the *Timaeus*. Plato himself put them together in that dialogue in a way which no idealist could accept nor any realist reject, for it seems to grant equal reality to forms and material things.

THE TWO-TIERED UNIVERSE

THE INTELLIGIBLE ORDER

Objects of pure mathematics
Laws of the experimental sciences
Empirically-verified universals

THE VISIBLE WORLD

Contents of the integrative levels in disorder
Properties of matter: energy-bundles
States of matter: gas, liquid, solid, plasma
Space: the extent of cover reference

Table V

The *Timaeus* contains the germ of a theory which supports modern thinking about the universe. At 52 B, there seems to be an indentification of matter with space. 'All that exists', he said there, 'must be in some place and occupy some space.' The earlier identification of space as the 'receptable' is well known (1). Plato added that 'in some perplexing and baffling way it partakes of the intelligible'.

Now let us look to modern physics and astronomy for further guidance. Studies of space are to be found chiefly in the discussions of cosmology. The curvature of space, according to Einstein's general theory of relativity and the red shift, is particularly relevant. The visible world is expanding, against local gravity, from a time zero of infinite density about 10,000 million years ago, at the same rate in all directions homogenously and isotropically, in symmetrical space with time asymmetry. The gravity is due to a deformation of the geometry of space-time around neighboring bodies.

That the universe as a whole has a structure must be the first thing to say; and that its structure is prior to the laws of nature the second thing. Unlike a part of the universe, of which there are many, there may be only one universe of a kind. It follows that the universe as a whole differs from any (or all) of its individual parts, for the whole cannot both function as a whole and also be introduced as an element among its various parts.

---

1   *Timaeus*, 51 A-B.

PART TWO. HUMAN NATURE

CHAPTER VI

MAN: NEEDS AND DRIVES

1. Man As A Product of Artifacts

Now that we know what sort of universe man inhabits, we can take up the account of his own nature as we left it in chapter I. The older accounts will have to be radically revised in the light of recent developments. More specifically, we need to take another perspective and look at him as he is conditioned by the artificial environment he has made for himself. Actually, it is doubtful whether an individual could make any move that did not involve an artifact of some sort. Therefore I will not try to say everything important that can be said about him, only that fraction which concerns his interactions with artifacts, which is, however, considerable.

In the present chapter, then, I will propose a theory of man based on the way his development has been conditioned by technology. I will suggest also a theory of human origins, a theory of motivation, and a theory of culture. And because all this makes life seem more orderly than it actually is, I will introduce in a following chapter the topic of aberrant man and his sheer perversity. I hope to show also that he does not exist in a void but in an environment, moreover one which is largely the product of his own efforts. Most theorists from Nietzsche to Heidegger have overlooked the fact that men do not merely 'behave', they behave *about* something, and that something consists in tool-making and tool-using. Therefore an understanding of the origins and development of technology holds the key to the understanding of human nature.

But I plan to go further than that. My thesis will be that while, as everyone knows, without man there would have been no tools, it is also true — and less well known — that without tools there would have been no man.

Man himself is a loose democracy presided over by consciousness, a sensitivity-reactivity system of organ-specific needs interacting with the available environment at various energy-levels: physical, chemical, biological, psychological and cultural, chiefly by means of artifacts. According to the archaeologists and vertebrate paleontologists, tool-using hominids are some 3 million years old, while, as we noted earlier, man in his present constitution goes back no further than 40,000 years.

I have already defined technology as the name for the invention and employment of artifacts, and artifacts as materials altered through human agency for human uses. Another name for artifact is tool; everything made and used is a tool in this broad sense: a wooden club as well as a skyscraper, a hand axe as much as a bulldozer. It should not be necessary to point out again that language consists in signs conveyed by means of scratches on hard surfaces or by shaped sounds, and so is also a tool.

Artifacts have been associated with the human species from its very inception. Society has always consisted in some kind of special organization together with its selection of tools, and this is no less true for a small primitive culture than it is for a great civilization.

Artifacts were not a new thing when man first learned how to flake stones and shape spear points. Some of the other animals had already invented artifacts: beavers had constructed dams, birds had built nests; but with those animals each generation had to begin all over again and there was no cumulative collection as there was with man. This marked a crucial advance.

Man is the only animal that did not passively adapt to the environment but instead set out to change it. Other animals, it is true, have made and used tools, but only man has passed them on cumulatively to successive generations. This is the crucial element in the human picture. There is reason to believe that this kind of 'tool use led to the distinction between human and ape ancestors' (1), that 'documentation exists for the succeeding Acheulean Industrial Complex, which was present in southern Africa almost certainly before 1 million years ago, and persisted with modifications probably until sometime between 300,000 and 130,000 years ago' for 'it is know that Acheulean people made handaxes, cleavers, and other stone tools' (2). Man first appeared because the use and inheritance of tools altered the anatomy of a pre-human species.

As we saw in chapter I, ever since the first man, or his predecessor in a previous species, hit upon the device of getting one piece of matter to alter another piece in ways which he himself could not do bare-handed, cutting meat with a bone knife, for example, man has taken a different road from that of the other animals. And when he passed on that knife to his son and grandson, with verbal instructions about the best way to use it, he took an even more momentous step, for it meant the beginning of an external inheritance.

1 Gina Bari Kolata, 'Human Evolution: Hominoids of the Miocene', in *Science*, 197, 244-245 (1977).
2 Richard G. Klein, 'The Ecology of Early Man in South Africa' in *Science*, 197, 115-126 (1977).

I cannot emphasize enough that what was distinctive about man was his inheritance of tools. This compelled an increase in the motor skills needed to manipulate them, and resulted in a corresponding increase in the size and multiple convolutions of the brain. The leap forward in human intelligence and in the ability to manipulate the environment was linked in this way to technology. Artifacts as material objects are of course part of the physical environment, and this new kind of interchange with it was what specified the peculiarly human as distinct from the merely animal.

The most important post-Darwinian principle is that man makes over his environment and then has to adapt to it. In one crucial respect, then, man does not behave like the other animals, he behaves like man. The difference consists in one principal development: man builds his own artificial environment to the extent to which his limited powers allow, and then hands it on to successive generations. This no other animal does. Thus, contrary to Wilson's thesis, human behavior is not only genetically determined, it is also environmentally determined (1).

It was man's inability to cope with his environment in the first place that led him to hit on the dodge of changing it to suit himself by making and using tools, that is to say, by learning how to turn some of the environment against itself for his own benefit.

The first artifacts were crude affairs and there were few of them. Early men were nomads, compelled to follow the migrating herds on which they subsisted, and so they were not able to carry much in the way of equipment. Civilization began only when someone discovered that the herds could be fenced in and crops planted to feed them, for that made it possible to settle down, build permanent houses and accumulate artifacts. With the introduction of animal husbandry and agriculture a stable community was made possible for the first time.

The mechanism for this was of course provided by the invention of language. The information communicated by language had become general, that is to say, it could now be applied to any number of things and not just to those at hand. No one has recognized sufficiently that abstractions, those of philosophy and mathematics for instance, have provided an increased measure of control over the environment.

I have pointed out that language is a material tool like any other; it endowed man with an epigenetic inheritance to add to the genetic inheritance that all organisms experience, chiefly by means of writing. To this inheritance each generation added somewhat, and the cumulative effect had its own consequences. Human behavior has been altered to accommodate man to the tools and languanges he has devised.

---

1 *On Human Nature* (Cambridge 1978, Harvard University Press). See also *Insect Societies* (1971, and more particularly *Sociobiology*: The New Synthesis, 1975, both also from Harvard University Press).

Let me sum up this part of the argument by saying that if man exists at all it is the result of some hominid's adaptive response to the inheritance and use of artifacts. By their means early man was in a position to reduce his needs more efficiently. Just what were those needs and how did he act to reduce them?

2. The Theory of Organ-Specific Needs

It would appear that man possesses a set of organs each of which has its special requirements, and that the whole organism acts as an agent for reducing them. In a word, the needs are organ-specific. I am not here concerned with organs which function internally to maintain the organism, the heart for instance, nor with those organs whose external requirements call for no special actions: air for the lungs and solid ground to support the upright skeleton. I am concerned chiefly with the fact that to reduce most of the other needs, purposive behavior, or, to describe it in a better way, directed aggression, is necessary. Directed aggression may be defined as the actions taken by the human animal to alter by force a portion of his environment and so make an artifact to reduce some one of his needs.

The organ-specific needs fall naturally into two groups. The first group may be regarded as primary because its demands are importunate: their reduction cannot be postponed. The second group is important rather than importunate because their reduction *can* be postponed.

The primary group of importunate needs are those of the stomach which needs food, the tissues which need water, and the sexual organs which need intercourse. Pain is the signal when there is a deprivation; the muscles of an empty stomach contract, causing pain, and when a mouth is dry, it produces a similar effect. The frustration of sexual desires can also be painful.

The secondary group consists in the brain which needs information, the muscles which need work, and the skin which needs security. Since these are less familiar when considered as organs with needs, some further explanations are necessary.

We do not ordinarily think of information as a 'need' of the brain, yet it is. If the young brain is not supplied with a certain amount of knowledge it will not develop. Children who have been allowed to grow to puberty without being taught a language are usually unable to acquire one.

Similarly, muscles have need of work. They must be kept active, and this means that they are called on to act aggressively, to alter by force materials in the environment. The effects of such aggression can be neutral, as in sports; constructive, as in the building of cities, or destructive, as in war. There are good reasons sometimes to want the first without the second, muscle-tone without destruction, for instance, but thus far no one has suggested a way to obtain this selective result. For, unfortunately, destruction involves faster and more thorough need-reduction than construction, and employs more strenuous aggression. Cities take years to build but they can be burned down in a matter of hours.

The last of the three secondary needs is for security; and this one will be the least familiar, for no one has considered it as an organ-specific need of the skin, yet as I hope to show, it is. Security means of course opportunities for survival, and survival can be either immediate or ultimate.

Let us consider first immediate survival, or survival in the near future: it depends upon an unbroken skin. Remember that the skin is the largest organ by weight of the human body. The average adult has 18 square feet of skin weighing six pounds, and it offers a first line of defense. To injure an individual usually means to break through his skin; so long as that is intact he has preserved some measure of security. 'Save your skin' is the old saying.

So much for the picture of immediate survival, but there is also ultimate survival to be considered. Man shares with the other animals the urge to live through tomorrow and does what is necessary to insure that this will happen. But there the comparison ends; for man alone takes the next step, which is intended to make it possible for him to live on indefinitely. Ultimate survival, which means survival after death, can be pursued by the individual through his identification with the cosmic universe or its cause. Here again the skin plays an important part, and most of the world religions, which are undertaking to reduce the need for ultimate survival, recognize that fact. Primitive religions for instance practice contagious magic, while 'world' religions establish rituals involving touch; for pious Jews the laying on of hands, for Moslems kissing the black stone at Mecca, for many Christians extreme unction and other sacraments of the Roman Catholic Church.

There is one essential difference between the primary and secondary needs. The primary needs call for an almost daily reduction, while the reduction of the secondary needs is a much more far-flung project, though reminders of it do recur with planned periodic frequency in religious ceremonies. Immediate survival is a requirement of what may be called the short-range self, the effort to stay alive. Immediate security is guaranteed by a sufficient supply of food, water, shelter, etc. Ultimate survival is a requirement of the long-range self, of immortality. Long-range security is provided for most individuals through the reporductive preservation of the species.

Corresponding to each need there is of course a drive aimed at need-reduction. It is the drives which account for most of not all of human activity. For each need there is an appropriate activity aimed at reducing the need, usually by obtaining the kind of altered materials the need requires.

The sequential life of the human individual is of course an indivisible whole, not a scattered affair of separate needs. All of the needs compete among themselves for attention, but not all can be satisfied at one and the same time. Needs sometimes cooperate in a drive and sometimes conflict. And because organs are parts of the whole, the individual establishes a rank-order for the needs by employing a system of priorities in which each need is assigned its opportunity to motivate the activities of the entire organism.

It is a peculiarly human phenomenon that often drives persist long after the needs have been reduced. The sex drive has caused oriental potentates to collect thousands of women, and ambitious business men to acquire property measured in billions of dollars, both far in excess of any organic need-reductions. Indeed it is a characteristic of man that in every one of his efforts at need-reduction he endeavors to exceed himself. This accounts for what the Greeks recognized as man's outrageous behavior, but also, on the favorable side, it accounts also for much of civilization, as represented by the monuments of art and many of the achievements of religion and science.

It may be noted parenthetically that the emotions for the most part are frustrated needs, the effects of needs which failed somehow to reach their goal-objects. They are the responses made by the whole organism to the blocking of intentional behavior.

3. Society and the Organ-Artifact Relationship

If now we consider the individual with all of his needs together, we notice that to reduce them *all*, to satisfy his thirst, say, as well as his craving for ultimate survival, an individual would have to control *all* of his environment. This no individual can ever do, even though he tries to as much as he possibly can.

There is another, and crucial, factor that enters into all of his activities. Even if he were able to reduce most of his needs to a satisfactory degree, he knows they will recur, and so he must make some provision for continued need-reduction in the future. Fortunately he does not occupy the environment alone, he is only one among many and so he encounters others like himself who are in much the same predicament. He invented culture by learning to use and inherit tools, but he developed civilization when he first learned to help in making larger tools which could be used only by groups working together. Groups operate by means of either cooperation or competition; both involve aggressions of some sort, constructive or destructive.

The individual is soon submerged in these larger aims. To prevent mutual destruction by many individuals in pursuit of diverse need-reductions, all must be willing to cooperate, and this means long-range planning, with need-reductions of all sorts now considered mutually beneficial enterprises. Social organization as an undertaking has the effect of increasing the size and use of tools, in short of rendering technology systematic and hence more extensive.

The result is a society made up of a set of social institutions roughly corresponding to the needs. An institution comes into existence because a certain need common to many indviduals is recognized as recurrent. Since the responses tend to exceed the stimulus, a regular channel for need-reductions must be provided. There are lines of connection running from each individual's organic need to the corresponding social establishment, with individuals and artifacts brought together in a conventional manner to serve it. An institution is made up as much of tools and instruments as it is of the individuals who use them.

Generally speaking, single institutions serve single needs. For the primary needs there are service institutions: farming in the interest of hunger, water-systems to satisfy thirst, marriage arrangements for sex. For the secondary needs there are higher institutions: armies for activity, schools and libraries for information, churches for security.

This is a simplified account, of course, but it provides the main outlines. Special features exist at every level, as for instance the two kinds of professionals who are to be found in every institution: those responsible for its productive aims and those who maintain the institution itself: men concerned primarily with subject-matter or adminis-tration. The state is always the institution which dominates the entire system, but it is often under the influence of some other institution, such as the church in the Middle Ages or science today.

Civilization means externalization. With the field of institutions we have entered into an area of greater sophistication, chiefly the work of artifacts which are so large and complex they bring about a new kind of relationship with the organ-specific needs of the individual: externalizing the bodily functions in order to perform them with greater efficiency. What civilized man can do that his predecessors could not is to carry out tasks that were not possible to him when he was limited to his internal organs and a few simple tools. In other words, an organ-artifact circuit has come into existence through which it is possible to do better outside the organism what had formerly been done only on the inside.

A few examples are certainly called for, and they are not hard to come by. The stove is an external stomach: cooking is a form of pre-digestion which reduces intrac-table foods to assimilable form and makes possible the consumption of hard fibers which could not otherwise have been eaten. Libraries are external memory-banks: they contain more information than any single human brain could manage. Computers are external minds: they calculate faster than mathematicians, and manipulate abstractions with greater skill and accuracy. Motor cars and airplanes provide external movement more efficiently than legs. Telescopes and microscopes are external eyes, they extend vision enormously, and reveal aspects of the environment that otherwise would never have been recognized.

Civilization also means intensification. Man has learned to use tools as a form of deliberate self-conditioning. A violin is an instrument that will do certain desirable things to him and others if he does the correct things to it. The same claim can be made for books, perfume and thousands of other artifacts.

There has been such an enormous increase in the production of artifacts that now the whole of man's immediate environment is made up of them, so that he lives largely in a world he himself has helped to build. This development has its own often unforeseen consequences. No one has been able to anticipate what effects the invention of a new tool will have, for technology not only influences behavior but often determines it. If it is true that species survive by adapting to the selection pressures in their environment, then it is clear that man has to meet a new challenge; what will his built environment do to him? What kind of human nature will result from his eventual adaption to this new kind of surroundings? It is much too early to tell, of course. Civilization itself is less than 10,000 years old. The evolution of species proceeds slowly, and changes will take hundreds of thousands if not millions of years to produce.

Meanwhile there is the immediate danger that artifacts might get somewhat out of hand. To understand this point, let us look back a little. The behavior of primitive cultures remains the same from one generation to the next. It is variety of what may be called *stereotyped behavior*. The behavior of civilized man alters from time to time, often rapidly, as in the present, and we may call this response *adaptive behavior*. But now a new and still third type of behavior, which may be called *instigative behavior* threatens to emerge: the building of machines which run other machines without human aid, such as thermostats and stabilizers, and even more complex combinations, such as the gas pipeline switching centers of interstate utility systems. The early models for this kind of technology were the Turing machines, plans for the building of computers which it was hoped could be taught to reproduce themselves.

What will be the result in the future of adapting to an artificial environment and developing and controlling still more sophisticated machines? Genetic engineering makes it possible to alter man internally. Environmental engineering provides the opportunity to shape the artificial environment. Both will bring about changes in human nature.

What kind of animal would man like to become? What kind of human nature does he want to develop? The answers to these ethical questions are probably not decidable on the basis of our present categories of thought. Yet the pressure of practical living make it necessary to come to some conclusion about them. In short, the destiny of the future of mankind lies in the invention and management of the necessary technology.

It lies also in the proper management of ourselves. Here we have been a signal failure. *Since primitive times there has been no progress in motivation*, it was and still remains ambivalent and therefore self-defeating. Early man wanted both to help and hurt his fellows: he wanted to help the in-group, and for this he had the shaman with his incantations; and he wanted to hurt the out-group and for this he had the warrior with his bow-and-arrow. The aims of modern man are identical, with the in-group and the out-group now represented by the vast populations of nations with their ever more complex tools. The shaman has been replaced by the doctor with his hospital, the bow-and arrow by the thermonuclear device. These represents enormous degrees of progress in technology, but the ambivalence of motivation is the same.

It was and continues to remain the most serious obstacle to the improvement of the lot of mankind, because it is a kind of built-in self-defeating principle. The last manifestation is perhaps the one that offers the most promising benefit as well as posing the most serious threat. The discovery of nuclear energy to replace the fossil fuels, coal and oil, is offset by the discovery of nuclear weapons. Add to this the ordinary cultural differences and it becomes clear that either mankind must recognize only those which promote no antagonisms or he is in danger of perishing from the earth.

4. The Making of Cultural Differences

Man is a polytypic species, but social barriers in no wise constitute isolating mechanisms; all men belong to the same species in virtue of the existence of a set of inter-communicating gene pools having biological continuity. Cultures are the results of diverse responses to varying conditions, and although not biological they are nevertheless profound. All men, it is true, have the same needs, but not all live in the same environment, and methods of need-reductions are perforce designed to accord with the differences. Surely no one would suppose that the environment of an Alaskan Eskimo was the same as that of a Kenyan Watusi. Despite the wide diversity of genetic variation, the resulting cultures operate effectively to impose similarities on individuals.

Culture may be defined as the local rearrangement by man of his available environment, it is the organization of men and such artifacts as they have been able to make out of their peculiar natural surroundings. Cultures operate by employing technologies suitable to the available materials in order to bring out their desirable potentialities, making certain of the possibilities of materials by transforming them. It is a characteristic of a high culture that it can make the most out of the least amount of material by employing the greatest amount of skill. How much can be accomplished like this is illustrated by the use of ink marks on paper employed as directions for how to use an old construction of wood and cat gut, as happens when a Menuhin plays a Mozart composition on a Stradivarius violin.

The building of a specialized environment by industrial man has led some authorities to suppose that he has been set free in such a way that his own development is unimpeded; but quite the opposite is the case, for industrial man is more tightly than ever integrated into his environment. Early man might have been able to survive in a natural environment, even in one quite different from his own, because his requirements were so few and so primitive; but modern man requires all of the culture he has constructed and would not survive very long without it. He has bound himself into a peculiar kind of ecosystem. Thus, far from being independent of his special environment, he is more than ever dependent upon it.

What is true of tools in general is true also of those particular tools we call languages. Because of the irrefrangable general nature of language, words which at first were combined only into sentences end by composing entire systems of ideas. Thus it is possible to explain the origins of abstractions which come to be exploited for their own purposes, chiefly in the explanatory schemes of philosophy, science and religion.

The danger is that ideas, which were designed to aid men in dominating their environment by making social organizations possible, end by exercising a tyranny over their thoughts, feelings and actions. Men who find themselves in the grip of an absolute belief will slaughter other men without the slightest compunction, as indeed they have done for millennia, and as fascists, communists and religious fanatics are still doing at the present time. Hitler's Nazis demonstrated what the fascists could do in this direction, the Pol Pot regime in Cambodia is a good example of what the communists are capable of, and the Ayatollah Khomeini shows how far the Moslem religious fanatic is willing to go. It is a sickening catalogue, from which no large group is altogether immune. Systems of ideas in conflict tend to drag men after them to kill or be killed wholesale.

There are local as well as regional cultural differences. The local differences are usually those defined by the boundaries and the pervasive influences of nation-states. The regional differences are wider and more elusive but when examined closely are found to be just as effective. There is a sharp distinction in this regard between European and Asian man. European man has sought control over his environment, Asian man has largely neglected the environment in order to seek control over himself. The former is outward bound, the latter inward bound. Both have pursued their goals in depth, which is to say in excess: both have sought to exceed themselves, the European in material constructions, the Asian in psychological transformations.

In sum, the world of material culture in which man finds himself immersed is as important to his development as his genetic inheritance, which turns to be the inherited part of the environmental influence which has survived through its effect on adaptation. By means of material culture he has been able to effect changes in his own existence which he could not otherwise have done. His need for survival for instance has been reduced through the extension of a life expectancy which has almost doubled in scientific-industrial countries. In this high adventure the individual is inevitably immersed. The social and cultural milieu is more than any one individual could surmount or even encompass. He can only participate in it to the best of his capacity, and perhaps, if he is that rare specimen, a productive and original individual, influence it a little.

CHAPTER VII

MAN: PERVERSITY

The ambivalence of motivation and the problems of cultural differentiation were the notes on which we left the last chapter. They are by no means the only difficulties in human life. There are many other conflicts and contradictions. I will choose two for consideration: perversity, and the effects of technology, factors found everywhere though not usually associated. Both are intensified and their moments of conflict sharpened as civilization advances. No examination of entire man would be complete without a consideration of what is involved when artifacts compel conformity and meet resistance at its outer edges.

1. Perversity

Perversity though common enough is certainly a neglected feature of behavior. Human nature has two sides, the one side making for order, the other for disorder. The orderly features of human nature are its most prominent. It is the one most frequently encountered as we live out our lives in society; yet there is another aspect, and it is seldom discussed. We must look at it here. I have in mind that feature of human nature which sets it in opposition to order.

Perversity is a word used to describe the wilful irregularities in human nature. It is the way in which the individual moves against order; sometimes against a particular order to which he may have taken exception, but more often against order itself. To be perverse is to depart from what is conventional in the ordinary course of things, to be obstinate, self-willed, stubborn and contrary.

All of us have those perverse moments when we refuse to cooperate or accept conventions. What I have in mind, however, is stronger than that: the perverse individual who rebels against an establishment and refuses to do what seems expected just because it *is* expected.

Every individual feels that from his own point of view he was born without being consulted into a world which is not of his own devising and is rarely suited to his tastes. No wonder, then, if he objects to certain of its features when he comes to full maturity or perhaps at that stage rejects the whole thing out of hand. He is, as we say, in a way entitled. For he is required to conform: the laws of the universe, if not those of his society, do represent to the individual 'legislation without representation', and he has a perfect right to protest. If he remains in it, however, he must in the end conform to it, and unless he commits suicide (as some in fact do) he must accept it to a large extent. Beyond that he may from time to time wish to reassert himself through acts of perversity.

By perversity I do not mean anything connected with mental illness, no degeneration or deterioration of the fundamental needs or tendencies, as for instance occurs in sexual sadism or fetishism. I do not have in mind anything abnormal or pathological. The treatment of perversity here assumes that everyone is perverse to some extent and for quite familiar reasons. Perversity by contract is the normal characteristic of the average individual. It is closer in meaning to 'ornery', 'stubborn', 'contrary', 'quixotic'. A perverse act is simply the opposite of what is normally expected.

There are, however, individuals who are characteristically perverse. The perverse individual is a rebel. The word has long political associations, but I mean something much broader than politics. Individuals in political rebellion usually act in concert, but I refer to something more unique and less social. In an age when even literature and the arts are conscripted to carry political messages, it is apt to be forgotten that not everything is political. The human spirit, that dominant inner quality of a individual which is displayed for the most part (though not exclusively) by his consciousness of his highest feelings, has little if anything to do with politics.

Perversity comes in many different shapes and sizes. It can be an occasional quirk or it can mark a prevailing life style, it can issue as an impulsive response or as a consistent behavior pattern. From separate individuals, it is only logical to expect contrasting tendencies, for no two are the same. The special features of perversity, those which make it typical, are to be found in impulsive actions, the taking of an opposite course for seemingly no reason at all. *Not* doing something just because it *is* expected is the archetype of the perverse act.

When an individual speaks in one way and acts in another, we excuse the inconsistency by saying that it is 'human nature'. That is hardly an explanation, but the difficulty can be explained. All individuals are pulled in two ways because they must deal with material realities while aiming at the ideal. The effort to reshape the world a little closer to the notion of how they think it ought to be brings them into abrupt contact with things as they are. And so they encounter brute facts which may not be to their liking while they keep working in quieter ways toward chosen goals. The division tears them apart and is responsible for feelings of frustration, often expressed as anger. Hatred then restores their right to assert themselves above and beyond the failure the anger first represented.

An important pair of conflicting interests is the desire on the one hand to belong to a social group and the desire on the other to be a unique individual.

The former is the simplest of the two and requires no special effort, everyone falls into it in virtue of birth and upbringing. A conformist is one who has chosen the path of least resistance.

The desire to be a unique individual presents a quite different challenge. It too may be a simple affair, a result of genetic differences which display themselves through maturation. But equally it may not, for the individual may rebel against conformity when he finds it militating against his desire to be different. The unique individual, the one who 'stands out from the crowd', may do odd things in an effort to find and establish himself.

The most familiar expression of perversity is the individual putting obstacles in his own path, which he does, often at the last moment, as when brides have been left at the altar, telephone calls not made, and promises broken. This kind of behavior is common and there is an explanation: by asserting his uniqueness, the individual thinks to make a last ditch refusal to surrender the self.

Perversity is indispensable. It is necessary for the individual in search of himself to act 'perversely'. Thus he may move — quite literally — against his own interests, in a way which seems strange and unaccountable to others and yet be quite logical. To behave in the opposite way from what others expect establishes him as an individual by compelling them to deal with him separately. Here is where all the rules that ordinarily apply to the understanding of human nature break down and a different, and additional, accounting is required.

Those who cannot through their own efforts share in the advantages of the social group to which they belong may act in such a way as to bring others down to their own level, and by this hope to achieve another kind of sharing: that of the has-nots. Such motives are unconsious of course, and the actions taken on impulse; but they serve to eliminate what cannot be attained in favor of a common denominator.

Vandalism may be defined as the deliberate destruction of artifacts without hope of personal gain. 'If I cannot have the good things of life, too, then I will destroy yours and that will at least, make us equal', the vandal seems to be saying. When young toughs break into a residence and destroy its contents, when a dish washer who has been refused work in a hotel burns it down, when people on the streets are attacked simply because they seem prosperous, these are acts of vandalism.

Vandalism is for the most part a city product: the more concentrated the population and the more advanced the technology, the greater the incidence of vandalism. It may be described therefore as a side effect of a civilization in which not everyone can fit. Those who cannot find their place express their frustrations in ways which tend to be random rather than planned, off-hand rather than deliberate. For everyone is compelled by nature to act.

Aggression, or the forceful alteration of material objects in the nearby environment, is a deep need of the musculature. It can take either of two forms: constructive or destructive. The constructive form is responsible for many material achievements, for all of the arts and sciences, in fact, for cities and their civilizations. But that is a slow process of need-reduction, and often something swifter is demanded. It takes centuries to build a capital city, but only hours to burn one down.

For the muscles of the solitary individual, destructive aggression is more efficient than constructive. Thus the vandal does after all get something out of his actions, he gets the reduction of his need for violent activity, a need which lies deep in everyone but is usually suppressed.

2. The Effects of Technology

Perversity may be seen as a reaffirmation of separate being. The effect of perversity is to make individuals different, the effect of technology is to make them similar. Thus technology has the inadvertent side effect of irritating the individual by aggravating the demand for conformity. The more advanced the society, which means the more complex and abundant its tools, the greater the degree of perversity in the individual. Thus he has the problem of coping with an artificial environment while endeavouring to preserve a separate self.

Technology intensifies the forces making for conformity at the time that they confront perversity with its strongest challenge. The products of technology are artifacts which by their very nature and function demand to be employed by everyone in exactly the same way. Driving a car, playing a television set, or using a telephone, does not bring out the uniqueness in the individual but to the contrary presses for conformity, and because its workings lie beyond the average man's understanding it is demoralizing. Tools and languages act to shape people uniformly and so operate in favor of conservatism. Artifacts often function as selection pressures. Because most manufactured articles are produced by the thousands and even millions in identical copies, they set the conditions for conventional behavior. There is no variety in the demands made by the electric toaster, the typewriter or the adding machine. All are inclined to reinforce a kind of destructive conformity and so tend to make robots out of men.

Just here is to be found the source of rebellion, which is directed not against the system that produced these conveniences but against the similarity of the types of behavior provoked by them. And this can be expressed in little ways, by small actions or passing expressions which seem to others perverse because they are not found to contain any justification. In social life, individual causes and effects may be separated in time so that they seem to have no connection even though it was there all the while but off to one side, so to speak, where it operated intermittently.

The more industrial life becomes mechanized the more individuals try to beat the game. No one wants to be reduced to a statistic in a survey. Computers present their own special challenge: the endeavor to handle masses of people sometimes meets with a resourceful kind of resistance. For the individual cannot abide being a number, and so he looks for ways to defeat the system.

The rise of the great populations in the twentieth century made it necessary for societies to operate in a uniform manner. No government or any other large institution, for that matter, can deal with that many people on an individual basis but has to lay down rules and procedures which everyone is expected to follow. This means that the individual is lost in the crowd, submerged in the masses so completely that his very existence becomes a mere digit. If I said that there was something special about the holder of a social security number, that in fact 436-20-1497 had a special contribution to make to society, you could hardly get the system as such to agree.

Civilization has surrounded the individual with a complex and multi-faceted environment. It is a challenge most individuals by training of various sorts are prepared to meet. Provided they are capable of acquiring a sufficient amount of knowledge of applied science and mathematics, they can learn how to design tools and even entire chemical and electrical engineering industries. If not, they can still perform single tasks, everything from pressing buttons and pulling switches to shorthand and typewriting. On the humanistic side, they can become journalists, television commentators, or writers; they may become painters or actors. Or they can teach and instruct the young. In a complex technological society there is room even for the individual who can perform only simple tasks on the belt line in a factory. Almost everyone, in fact, can find his niche in a public world where all things must work somehow together.

There always have been born losers in every society, however, those who could not measure up to its demands, and the demands in a technologically advanced society have multiplied many fold. Much is called out in the way of intelligence and manual skills; and so, granted also the increase in population, we should expect that there would be many failures.

The lesson of this chapter is that, given the demand for conformity that is built into every society to the extent to which it is an organization at all, every individual in it is to some small extent a misfit who seeks to express his individual difference in some way, either in harmless forms of expression or in aggressive and violent acts of destruction.

3. Disorder and Irrationality

Perversity is a class property. It is as genuine a manifestation of human existence as anything else, and has effects which cannot be overlooked. Since reality as such is characterized by equality of being, it must include everything on an equal footing; it cannot be a matter of selected preferences, and so the rejected side of human existence: crime, rebellion, ill will, must also be considered. The real cannot be limited to the rational; not certainly if we wish to count quality, for then we must include also the quality of force, and physical force is no respecter of moral priorities.

It comes to this, then, that we need a new conception of reason. What does not answer to law must still answer to reason, which is the larger category. Disorder is a product of the intersections of many actual orders. Men live together in a common culture which they have woven together and which in a sense reorganizes them. Its essence, which is its type of order lies at its center though often remaining hidden, implicit, assumed and covert. That all orders are limited and therefore disorder more prevalent is a fact of existence which has been sadly overlooked in every ontology. Disorder, irregularity, fractals, irrationality, surds, all have to be accounted for, unless reason is to remain incomplete.

Each individual finds himself living as a particular in a domain of particulars. What he has acquired from previous encounters is the knowledge that all particulars are the elements of order — but mostly out of order. The so-called 'order' of experience is not truly an order, only a succession. For instance the observer in his ordinary experience may encounter a roach, then a man and next an ape, but biological evolution calls for the roach to have appeared before the ape and the ape before the man. Or the observer may encounter five trees, then three shrubs and finally four hedges, although he would know from prior studies that the logical order of the numbers themselves is 3, 4, 5. Indeed the domain of universals was discovered by suspecting that there is a proper order of elements behind every instance of their discovery in the domain of material particulars, an order which can be isolated abstractly.

All material particulars are engaged continually in interacting with all others by means of force. Everything actual affects and is affected by everything else actual, in a scene of unremitting activity. The domain of universals, on the other hand, which is a domain of essences, stands unchanged forever as a permanent set of possibilities. Given the right conditions in existence, there could always be a set of four things, just as there could always be a roach, an ape, a man. There is a mixed-up-edness about existence, which is filled with turmoil but also interwoven by a myriad of separate aims.

Perversity, as a matter of fact, is the hallmark of striving and turns up in every corner of the world occupied by man. What the world does not do to the individual he usually does to himself. Men have conflicting desires, and act accordingly. The individual always struggles on two fronts, within himself and in society. The lessons of life for most individuals mean watching helplessly while their chosen goals recede despite all their efforts, because through their own machinations as well as through circumstances they have become trapped in a series of accidental frustrations.

Like any other animal, the human individual acts on the basis of responses to stimuli, but his actions are also the outcome of feelings and of chains of deliberate reasonings, of habits and impulses. It simply happens that all of his activities cannot be coordinated; there is no integral whole to be made out of the sum.

Now look again at the world in which he acts. It, too, is the result of many cross currents: natural events which might interfere, such as tornadoes; social events such as revolutions, as well as far more numerous little versions of such events, such as winter storms and street riots, respectively. He must deal with other individuals separately and in the round; and with social groups and institutions, most often arbitrarily.

Of course technology plays its part in the social context, too. Here there is a time lag and an incongruity involved. How long will we continue to express ourselves in terms of an ancient and outmoded technology? Does it still make sense to talk about 'where the axe will fall', to 'put the cart before the horse' or to 'sail close to the wind', in a day when axes, horses and sailboats are rareties or at best playthings? I should think it would be better to talk about 'where the bulldozer will cut', 'put the trailer before the truck' or 'come down safely in a crowded airport'.

Beliefs and customs — profound ideas and ordinary usages — all tend to lag behind the developments in culture which affect everyone. People still throw salt over their shoulders when they spill something, they still avoid walking under a ladder, and they still refuse to live on the 13th floor, all while denying the beliefs these actions imply. Worst of all, perhaps is the current vogue for astrology, which has long been totally discredited.

CHAPTER VIII

MIND: PERCEPTION

1. Consciousness

'Mind' is a very loose term, one that has been used over the centuries to cover a vast number of conceptions and misconceptions. Here I mean to add one more description, this time aided by recent findings in the sciences, particularly the science of neurophysiology.

I pointed out in the last chapter that the question of what man is can no longer be separated from the prior question of what the world is like. Mind, I said there, is now considered to be a function of the brain, and the brain of course belongs to material world. Kant worked on the hypothesis that there was a necessary relation between man and the world, and in so doing he initiated the theory of knowledge as a formal study. Ever since his *Critique of Pure Reason* was published, it has been interpreted as a conception of the world immersed in the knowing mind. The present study reverses this interpretation, and endeavors to present the outlines of a theory in which the knowing mind is immersed in the world.

Mind, it is now believed in some scientific quarters, is the name for the brain functioning, and the brain is the most complex form that matter takes. Its emergent quality (consciousness) depends upon such an intricate structure that it is easily destroyed. But that does not challenge the nature of its material underpinning: matter increasingly organized from quarks and atoms to cells and organs.

The success of the physicists in exploring the material world left to the pholosophers their old refuge of the private world of consciousness, the self and the mind generally. By exploring this world they succeeded in scoring some notable gains, but much yet remains to be done. For the neurophysiologists have shown that the individual together with his private thoughts and feelings was never a valid isolate. His external relations and his dependence on external objects are still to be accounted for.

Philosophy, one tends to forget, was never science-free. Philosophers identified their work with the older scientific findings so firmly that they tended to defend them against the new findings, on the mistaken assumption that they were defending philosophy against science. But genuine philosophy always has been open to new knowledge from whatever quarter.

Lately it has had to face fresh information concerning its last private refuge, which is the consciousness itself. Consciousness is a fact that no one would wish to deny; and it is also a fact that any quality as such is essential simple and indescribable, as we have noted already in chapter IV. But what the fact of consciousness represents is open to inquiry, one to which the sciences have contributed.

From the experiments of the neurophysiologists we learn that consciousness involves a two-way dependence: it depends upon novelty of input, and upon a high degree of alertness supplied by the neural centers (1). The more sensitive the organism, the more helpless it is in the hands of the data. Specifically, consciousness is the quality which arises when there is a state of tension between a higher nervous system and an available environment.

Those who, like Husserl, place great emphasis upon consciousness usually mean not consciousness but its contents. When the early knowledge of nature seemed to disclose a world of simple matter and a close-in universe, it did seem as though the mind with its faculty of imagination was the richest area presented for study. The rapid advance of the physical sciences has changed all that. Now the universe stands exposed as an exceedingly complex affair at all levels, so that the imagination has been easily outpaced by the facts.

Consciousness itself may be here defined as the qualitative correlate of controlled perception. Since all of the mental life is tinged with quality, the quality of the mental life is indescribable, also. To the extent to which qualities enter into experience (and it is not total) this is discouraging. Not all of the mental life is qualitative, however. There are frameworks for it, and to the extent to which they exist we may retain our hope of describing something about it.

There is little to be learned about consciousness beyond what the neurophysiologists have said, namely, that it is the state of alertness or wakefulness of the organism. Mental states are qualitative emergences of states of the central nervous system, in this case the functioning of the reticular activating system, the mass of neurons near the top of the brain-stem reported by Magoun and his colleagues (2).

1 W. Grey Walter, *The Living Brain* (London 1957, Duckworth).
2 H.W. Magoun, *The Waking Brain* (Springfield, Ill., 1958, Charles C. Thomas).

Let me hastily add the qualification that insofar as it is consciousness we mean when we talk about the mind, this information implies no reduction of the mind to the brain. The mind is dependent upon the brain but the brain is not the mind. The mind is rather the way in which the brain works. As C.H. Waddington puts it, 'intelligence is seated in the brain and the brain is part of the body' (1). That does *not* equate intelligence with the brain, however. The laws of the levels of organization are involved here: and the higher cannot be reduced to the lower however strong its dependence. The existence of emergent qualities warns us away from any such interpretation: organizations are always one level higher than their elements of analysis.

Consciousness depends upon the high frequency of nerve impulses but cannot be reduced to them. The brain is an electric machine which runs on 25 watts (2) and makes consciousness possible. Consciousness as such a non-reflective, it deals with the object, not with itself. It continually receives signals from the outer world or from the body, both of which of course are external to it.

The first step in consciousness comes from the force of the object impinging on the subject. Peirce was at great pains to argue that the knowing subject is helpless in the hands of the data. He was compelled to acknowledge what seemed to him to be true on the basis of reason and fact. The evidence of novelties and surprises would seem to indicate that consciousness does not determine its object but rather the reverse. What Brentano, and Chisholm after him, calls the 'intentional consciousness' is in effect a second stage; it might be another name for attention, which is often deliberate. But that does not alter the nature of the first stage, which is not. In the first stage of knowing there are no deliberately chosen targets. So long as an individual's eyes are open he is powerless not to see what stands before them. Once faced in a given direction he is compelled visually by what lies there.

Chisholm himself in a later work in which he appeals to animal faith and to the role of a synthetic *a priori* (3) seems to have recognized the inadequacy of the intentional consciousness. Consciousness is free to choose its objects but only within certain limits. Such limits are determined by the state of the organism. It happens for instance with some regularity that pressures in bladder or rectum preempt conscious attention under appropriate circumstances, and there are many other such organic preemptions. Usually even if not always the direction of consciousness is dictated by some organic need, and the consequent drive which is initiated at the conscious level is not merely a passive intention but rather a directed aggression, which to be effective nearly always involves

1   *Nature*, 281, 326 (1979).
2   W. Grey Waleter, *op. cit.*, p. 75.
3   For the 'intentional consciousness' see Chisholm's 'Intentionality and The Theory of Signs' in *Philosophical Studies*, III 1952, pp. 56-63. The later work is *Perceiving* (Ithaca, N.Y., 1957, Cornell University Press).

recourse to some instrument. Consciousness mediates between the specific need and the object needed, but it is impossible to have directed aggression without conscious intention.

There are many kinds of nervous responses; consciousness is only one kind, a response at the highest level of the central nervous system. The organism responds to many stimuli in a way which does not affect consciousness. An individual may die from inhaling carbon monoxide without ever feeling or knowing what is happening to him. Some types of carcinoma give no pain signals until the very end. It is possible to fall in love without recognizing it for some time. And, above all, the unconscious processes of thought take place in such a way that the individual is not aware of them until a conclusion has been reached.

Another way to describe consciousness is to say that it is the state of the subject when that subject is turned toward an object. The object may be a material object out in the world, or an internal object, such as an image which we say is 'in the mind'; but without an object there is no consciousness. In the uncharted alembic of consciousness, fragments of physical and social events take on a strange arrangement, but all of them still come originally from the external world.

As a state of wakefulness, consciousness is hopelessly dependent upon that world. The neurophysiologists have confirmed experimentally that monotony of input causes the subject to lose consciousness, as may happen with a driver of a car alone on a long straight road at night. There must be a varied input if the subject is to remain alert. Thus consciousness is a dependent function, the arguments of which are the material objects in the external world considered as variables.

Consciousness is the internal end of perception. It is dependent on the tensions between the brain and the world, involving communication across the brain when operating as thought. As a function it makes a connection with the world by means of the central nervous system, and may be considered in isolation only with the prior understanding of its dependence. Introspection is the awareness by the subject of the fact that he can now operate by means of signs by finding relations between them.

I have reserved all mention of the 'self' until after the discussion of consciousness because what we mean by the 'self' these days, at least in philosophy and psychology, is the conscious or aware subject. When a subject is conscious of his own state of consciousness, it is called self-consciousness. But self-consciousness is the consciousness of the consciousness of objects, never the consciousness of self alone. Since Hume it had been clear that there is no special location for the self and no special composition has been discovered for it. No doubt those reactions which we attribute to the self take place somewhere in the central nervous system or the cortex, but just what this means or how it operates is not well understood.

All that we can say with safety is that there is a consciousness and that there is something which is conscious. To ascribe the latter to the human spirit is no help unless what is meant by spirit becomes better known. If it means what I have said that I mean by it, namely, the dominant inner quality of a human subject, then we are only a little wiser. For consciousness as a fact refers to a function, and there must be something which functions.

The common understanding is that the self is a synonym for the knower. Perhaps; certainly knowing is part of the work of the self, but that is not the whole story because the self is not only that which knows but also that which senses and therefore cannot be identified with the knower but only with the knower in one of his functions. As well as knowing, the self experiences pleasure and pain and is able to receive the slightest trace of sensation. Until we achieve a greater insight we shall have to consider the self as the name for the center of all experiences, including consciousness.

That the self is a center implies that it is the center *of* something. The 'something' in question is the rest of the organism. The human individual is a moving, sensing, vibrant being with an organization, and that organization is centered on the self. The unity of the organization with the self at its center is always a tenuous affair. For, as the psychologists have learned, everyone is at least two people, though they tend to merge in most· cases. However, 'leading a double life' is not all that uncommon, and in pathological cases there may be two distinct personalities involved in the same person, with one almost unknown to the other.

2. Perception as Encounter

I shall try now to examine as naively as possible the kind of happening that we have come to describe as 'having a sense experience'. In this way I hope to set up a first stage in the analysis of knowledge. The knowledge relation itself is a delicate one, so easily disturbed by close examination that the results are apt to be misleading. Just as in physics, according to the principle of indeterminacy, light interferes with the study of photons, so in the analysis of knowing self-consciousness distorts the accuracy of description of the awareness of the external world. The process operates in much the same way that getting control of the autonomic functions distorts them. If one thinks abou breathing, the regular rate is upset and any conclusion one reaches about its frequency is almost sure to be false.

Every sense experience is an encounter between a human individual and a material object, and it is this encounter that I want to examine. What I have been calling the subject, I will now refer to as the observer. Let me quickly comment first on the observer and next on what he observes.

The term observer will be employed to describe the outward-faced aspect of the experiencing subject as he stands in some relation to a material object, but this is only another way of talking about a perspective. Now a perspective is something which belongs to an object and is carried about by it. The table seen from directly above will be the same wherever the table is. A perspective must be understood as a view made possible both by the angle of vision and with the equipment of the observer. What can be experienced from a particular perspective is the arrangement of the world as experienced from that angle.

The equipment of the observer is relevant, for it contributes to what can be experienced from the chosen perspective. Thus an artist standing in a valley from which he can see the rock face of the side of a mountain would not 'see' the same features as a geologist who was standing on the same spot. The difference is not exclusively mental, however, for what is true of the perspectives of the two men would be equally true of a camera with two kinds of film. Thus a camera with black and white film would not apprehend the same impression as another in the same position but with color-sensitive film. It is clear then that we are talking about aspects of a material object, not those of the observer, for what we are concerned with is not the object in relation to the subject: for the purposes of knowledge the dependence has shifted, so that our concern is with the subject in relation to the object.

So much, then, for the observer. As for the material objects he encounters, they are the things and events which are represented by static and dynamic states respectively. Since matter accounts for the resistance to muscular effort, the best indication of its presence is action.

I should add quickly that I have selected the most general case because in this way I do not have to give special consideration to the encounter when the material object in question is another human individual. The second individual, in this context, at least, is simply another material object. Whenever there is an encounter between two minds, there is also one between two material objects; for brains are parts of human individuals and themselves material objects.

More complexly, the encounter generally involves a third object, such as the sounds of a spoken language or the marks of a written one or some other species of artifact. This additional factor complicates the problem and require analysis at a higher level of organization. Most of the encounters which take place in perception are with artifacts of one kind or another, chiefly because that is how nearly everything within reach of the unaided senses can be described. After all, we do live in an almost totally artificial environment, at least, the available environment is, and few try to perceive beyond it. Thus it is fair to say that perception is conditioned by artifacts.

Before entering into a discussion of the details of the encounter of the subject with material objects, I had better talk about the subject's predicament in general. To take the point of view of a subject would seem at first glance a most easy and natural one, since we are all subjects. Unfortunately, this is not the case, True, we are subjects and so we have the point of view of subjects; but to *have* the point of view of subjects is not to *take* the point of view of subjects. To be aware that we are subjects and that, as subjects we have the point of view that goes with them is to take a second-level subjective view, not by any means an easy or natural thing to do. Such a second-level viewpoint involves a double subjectivity. It means not only that an observer sees a horse, for instance, but also that he sees himself seeing the horse. Such double subjectivity is rarely recognized in the literature of perception.

In the perspective of the second-level, there is the situation of a subject viewing both a subject and an object, and no less so because the subject viewing and the subject viewed are one and the same subject. Obviously, this involves a neglected assumption and it makes for concealed difficulties in most of the Kantian epistemologies from Cassirer to C.I. Lewis. The minute we put a man on the moon, Kant was done for.

In a realistic approach no such assumption is made, and that is why I have thought it best to call attention to it, and so I have named it the Method of Induced Subjectivity. It consists in deliberately remembering that the observer is a subject and that he falls naturally into the subjective point of view. As a matter of fact — and this is the important point — it means having a second observer capable of viewing the subject and the object of knowledge as though they were two different sorts of objects.

Subjectivity is a tricky affair, and it can be induced in several ways. Not many have noticed that there is a difference between the inner-induced and the outer-induced varieties, but the distinction is crucial.

The religious cultures of Indian Asia have been largely devoted to inner-induced subjective states, while the secular epistemologies of European countries have been occupied with the analysis of those which are outer-induced. The latter can be made eventually to give way to a concentration on the object toward which it was originally directed; that is to say, to the object as something having independent being.

The assumption underlying outer-induced subjective states is one which was frequently forgotten or overlooked in the kind of analysis of knowledge so popular in England in the seventeenth and eighteenth centuries: that in this context the subject is dependent upon the object and conditioned by it, and not the reverse. The dependence on the object is forgotten because the observer is always able to shut off further sense experience, to close his eyes, say, and concentrate on what he has already seen and which exists for him now as images.

Subjectivity is merely the subject mentally digesting the impressions from material objects he has received previously through his sense organs. The subjective procedure does not alter the fact of the origins of what he is experiencing: it came to him from the external world, and is not dependent for its being upon his knowing. There is, then, strictly speaking, no such thing as 'subjectivity'. What is *called* subjectivity in any one of its varieties, not excluding solipsism, is only some form of objectivity closed or open, and, in the case of extreme subjective states, usually closed.

When we act in accordance with the belief that the objects of our perceptions exist independently of those perceptions, more often than not the objects will respond accordingly; and such behavior on their part reinforces our belief about them. So that it is positive action rather than passive perception or equally passive thought on which we must rely in the last analysis to support our belief in the external world. Hume hinted at this and so did Reid, but because Kant was the more powerful commentator the point was passed over, and the efforts of the American pragmatists to show that activity is a source of reliable knowledge, despite the powerful evidence for it which could have been adduced from centuries of experimental science, did not receive their due from the philosophers.

3. Perceiving as Knowing

We shall be interested from here on only in outer-induced subjective states. Knowing, on this approach, appears as a special case of being, the being of being-known, or being from the outside. Knowing-as-such is being from the inside. For there is objectivity inherent in the very fact that the subject is able to recognize himself as a subject. The proper approach of philosophy therefore, is to take its stand on the object, not on the subject, and to consider a subject only as an object experienced against a background of other objects with which it is constantly interacting.

If then we are to take our orientations from the object, we would do better to choose a proper object and not the subject-as-object. The subject would be a poor choice of object for this purpose because of the passivity inherent in the knowledge process. That we do not perceive our perceptions or experience our experiences is, as Herbart saw long ago, a strong argument against being limited in our theory of knowledge to what we can perceive and experience. The observer does not see his eyes bu only what his eyes see, and the content of his experience is objective to both of them and indeed must be independent of them since it cannot be changed in this way. It is not without good reason that the physicist prefers his instruments to his body: relying upon the vibrating membrane rather than his own ear, attending to the thermometer rather than his own skin, and employing telescope and microscope rather than the unaided eye.

Evidently we shall be obliged to debit the observer with an inability to explain the content of experience altogether by means of the act of experience. It will be necessary to look elsewhere. It is true of course and always will be that the world as we know it is the world-as-known. But to suppose that the world-as-such is bigger than the world-as-known, as indeed it must if we are to admit that from time to time there take place important and sometimes surprising additions to our knowledge, has important consequences to the theory of knowledge. For the world-as-known is dependent upon the world-as-such, in the same way in which a part is always dependent upon the whole of which it is a part.

Knowing is an accidental, not a necessary, property of things known: they would exist if they were not known. Even mental images, which presumably would not exist if they were not known, have when they do exist relations to other and presumably nonmental things. Works of art after all are congeries of bits of real experiences assembled anew. Moreover, it can be asserted of all things known that knowing them does not produce them. That this proposition cannot be demonstrated as a matter of knowledge has been the main argument of the subjectivists. The counter-argument would seem obvious, however. For the proposition that what we do not know we cannot know is equally undemonstrable; it cannot be proved in the process of knowing. We may know that we know, but we do not know also that we produce our knowledge in the course of knowing, and there is no evidence of this sort available.

That things known stand in some relation to the knower would seem to be a trivial tautology. That the things known, however, have other relations is not open to the same charge. The chair as I see it stands in some relation to my visual perception, yes, but more importantly for the purposes of being a chair it has relations to the individual who sits on it, to other chairs with which it shares class-membership, to the floor which supports it, to the roof which keeps the rain and snow from destroying it, and so on. It is somewhere in this broad picture that the observer must fit himself and his knowledge.

The point is that what is peculiar about the observer is his ability to add to his experience of objects by relating himself to a different and quite separate kind of experience: the experience of objects encountering each other. Is it possible to deny that there are interactions between the things that the observer experiences? Can he rightfully claim that all things stand only, or even chiefly, in relation to him and not to each other? If the observer were to see a man and a horse, he might if he wished suppose that their existence was dependent upon his experience of them, even though this would seem to be stretching things a bit; but it would be even harder for him to continue supposing it when the man mounted the horse and rode off: was his experience responsible for this, too? To call such an event an experience is an admission of what the observer owes to its content.

What is always involved in the relation between the observer and the material object which he observes is that it is a function of two variables; both changing in time; the observer as a dependent variable, and the object observed as an independent variable.

I propose to call his observation a passive conditioned encounter. It is passive because the observer is helpless to alter his impressions as he receives them. Moreover the encounter is powerful and it is insistent. The observer cannot get away from it or avoid its impact upon him. It will have its way with him so long as he exposes his sense of sight to it. If he looks at the yellow full moon, he will be unable to see it as red or square. His activity has been responsible for the exposure: he has placed himself in the position of seeing the moon by looking up at the sky on a clear night. But then the situation immediately turned passive, and he lost all influence over the terms of his impression. The moon was yellow and round, and that was that.

The observer does tend to get his experience confused with the description of his experience. He does not describe his experience until there has been a recognition of the objects experienced. But if the objects are too large to be grasped in a single act of experience, the recognition may have to come later. Experience may be of simpler elements: a color, a shape or an odor, when he knows and conditions his experience upon knowing that there are no such things as colors, shapes and odors that are not the colors, shapes and odors of material objects. But he belongs less primitively with experiences which are describable; in his description the class of material objects is as primitive as the singular material objects themselves because it is the class which enables him to describe his experiences.

So far there is nothing either novel or notable about the account; it has been set forth many times. But I did introduce a new note, the one contained in the word, 'conditioned'. The passive encounter is conditioned by many factors. I shall choose the most important one. The object encountered is a member of a class.

The important thing to note about the general aspect may be understood best if we frame its description in terms of the dual aspects of perception. When an observer perceives a material object, he sees it simultaneously in two ways: as an individual thing and as the member of a class. He could not do the one unless at the same time he were able to do the other. Thus the recognition that the material object is the member of a class is necessary to perceiving it at all. Hamlet played with this idea satirically when badgering Polonius, in III, ii, 401-406.

Ham. Do you see yonder cloud that's almost in the shape of a camel?
Pol. By the mass, and 'tis like a camel, indeed.
Ham. Methinks it is like a weasel.
Pol. It is backed like a weasel.
Ham. Or like a whale?
Pol. Very like a whale.

If something were to exist that was unlike anything we had ever seen before, we would have difficulty in seeing it. But could that actually happen? Universals are perceived in things as their classes and relations. There is a genuine *perc*eption as well as *conc*eption of universals. I will have more to say about this in the next chapter.

4. The Contribution of Technology

Everything that I have said thus far in this chapter on the process of perception, and everything written about it by others, for that matter, applies to the mesocosm but not to its adjacent segments of the microcosm and the macrocosm respectively. The 'common sense' of perception is what the ordinary individual perceives without any artificial assistance. That certainly does represent the larger part of his experience, but at the same time is very far from being all of it. If you will recall what I said in chapter V, section 2, about the three-segmented universe, then you will recognize that unaided sense experience applies only to the central segment, though even there artificial aids are not unknown. I have only to cite the fact that most individuals eventually resort to the aid of corrective glasses to improve vision.

Admittedly, for most individuals in the course of their ordinary round of experience, the facts are as I and many other have stated them, and usually there is no occasion for them to expatiate on other areas which they conceivably might experience provided they were furnished with the requisite artificial aids. But even so that is very far from being the whole story.

For the perceptions of the other two segments of the three-segmented material universe, which have been named the microcosm and the macrocosm respectively, instruments are needed, and a special technology has been designed for them. With each new instrument we acquire more experience, and hence more knowledge, of the material universe. Everthing from the electron microscope to the Very Large Array of dish antennas contributes to our knowledge of the extended environment.

In the long run it seems inconceivable that any comprehensive theory of epistemology or of the psychology of perception could remain unaffected by the new knowledge of the three-segmented material universe. The relations between them in an unbroken chain certainly do make any one segment considered by itself as a separate affair seem misleading, and this is the case with the central segment which for centuries has been considered the only one. The changes in that situation with the discovery of the two adjacent segments calls for a radical revision of theories, and indeed one more extensive than I have undertaken here, although in the early chapters of this book I have endeavored to suggest from time to time some of the ground that will have to be covered.

CHAPTER IX

MIND: CONCEPTION

In the last chapter I tried to show what the new empirical findings had to say about the mind in terms of its most primitive and largely unassisted functions, consciousness and perception. Here I hope to elaborate the discussion by adding some reflections on mental states and the process of conception.

1. The Experiences of Entire Man

The observer considered as an entire man is capable of having three kinds of experiences. He can feel by means of his senses, any or all of them; he can engage in actions, either casual and unassisted or planned and assisted by artifacts designed for the purpose; and, finally, he can think, again either casually or in a calculated manner, also the former unassisted and the latter assisted by artifacts of logical or material provenance, say by rules of deduction or by a computer. For the observer, being present in a particular way, then, means being present to all of him, and this requires a sequence of events because the exposure of his senses cannot be accomplished all at once.

An observer whose single faculties operate separately, one at a time, does not exist. No one has ever depended entirely for his acquaintance with the external world upon a single sense, such as that of sight. The contacts come in terms of all three faculties interacting together. In a word, he deals with the world as though he were a whole individual, not a fragmented one. The senses, aided by action and thought, present the observer with an integrated conception of the world, and it is in terms of such a conception that he is able to conduct himself. His very success at living constitutes a kind of pragmatic reinforcement of his conception. The composite nature of his experience (for that is what every experience has) is organized into a synthetic unity by means of a stereoscopic representation which is largely true to the facts. He sees things in immediate perception at least partly as they are, and he supports this with a conception of how he knows them to be.

There is much more to be said about this aspect of happenings. I have deliberately omitted a number of subjective conditioning factors which are quite essential, such as the degree of alertness. We saw at the beginning of the last chapter that since wakefulness depends upon a level of cortical activity which has to be maintained by a continual stream of incoming sensory impulses, it would seem that consciousness depends upon the external world and not the reverse. Admittedly, there must be a capacity for receiving and registering these impulses, but the capacity could be there without being exercised. The existence of consciousness, then, can hardly be adduced as evidence that the mind has a mode of being which is separate from the material organism. There is also an element of abstraction involved in attending to a material object to which it essentially belongs and with which it can be apprehended in a larger segment of experience encompassing a wider field.

Resemblances recur, differences perish. The way in which material things resemble each other is often encountered again, but the ways in which they differ will be lost and their places taken by other differences. Substance *supports* the similarities but *contributes* the differences. These are the facts the observer has to face in the data disclosed to him by his sense experience; and they are facts so insistent that he is compelled to recognize a recurrent order of similarities and a persistent sequence of differences. Similarities provide continuities while differences are responsible for discontinuities; but the similarities and differences can be combined, and indeed often are, as can be observed in any cycle: the seasons, for example, or the face of a clock. The differences are saltatory; as it happens their abruptness furnished the discreteness necessary for discrimination in the first place. That is why continuities have come to be regarded as unobserved even though they too are observed.

I noted that an object present can be considered from two aspects: the one particular, the other general; and now I want to concentrate more importantly for the moment on the more general aspect. My example, if you recall, was the presence of the moon. The important fact about the general aspect is that the other members of the class to which it belongs are absent. They are absent in time and they are absent in space.

Planets are absent in time because even though they last for a great while it is not forever. No one knows the exact age of the material universe but everyone who has studied the question agrees that it has been here for a very long time, some 13 billion years, at the very least. In all that vast duration there have been many planets, most of which no longer exist but all of which will have to be counted as members of the same class even though they are not present.

Planets are absent in space because the material universe is an exceedingly large one.

A million galaxies are known to exist, and how many more it is impossible to guess. As far out as the telescopes reach, galaxies are still observed. Thus we may say that in space as well as in time the absent planets are to be found in very great numbers, so many that it is almost impossible even to imagine them.

How do these facts bear on my example of the encounter of the observer with the moon? It raises the question at least of whether what is encountered is a unique individual or the member of a class. The moon is a unique individual, no doubt of that. It has properties which it shares with no other planet. It has features which are peculiar to it and it occupies a spatio-temporal region occupied by no other body. Yet it must be admitted that its uniqueness, though it is a fact, is overwhelmed by the further fact of its membership in a class having very many members. It is the sole *present* representative of the class, and this is so meaningful that it tends to overwhelm every other aspect of its existence.

It is at this point that I find it necessary to take up again the importance of the class as an element present in every encounter, this time of the observer with the moon, by making two moves. One is to acknowledge the operation of a kind of Principle of Class Presence. The effectiveness of a class is a characteristic which will have to be included among the essential elements in any happening sufficiently formal to be called an encounter.

What the individual is meeting when he encounters a planet, or any other material individual, so far as that goes, is the member of a class, and one of the salient features of the encounter is the power the class has and imposes on him through its present representative. And because of the prevalence of members throughout the vast reaches of time and space the class is of the utmost importance, of more importance, in fact, than any member through which the observer may come to know it. Every individual exists in part as the only present representative of the class of which it is a member. So that knowledge is not merely particular but also general, and every experience contains elements whose representative nature is the most prominent thing about it.

This brings me to the second of the two moves. If the first move was the introduction of a Principle of Class Presence, the second is the recognition that every encounter is a primary confrontation.

In every act of experience there is evidently a complex situation. There is the encounter with the particular material object, and there is the confrontation with the objectivity of the class, with the representation in the present of absent objects. Here the three capacities, namely, for thought, feeling and action, which I have been at pains to show must be separated, though this time with a different intent. When the observer thinks about his own thoughts, this at its best is logic; when he feels keenly about his own feelings, this at its worst is neurotic; and when he acts he is changing his spatial relations with the objects about him in an effort to use the encounter to his own advantage.

The individual must recognize at some point that he has met with two distinct but related kinds of resistance. Not only is he powerless to render his experience with any particular other than it is, but he is equally powerless to alter the class to which it belongs. Logic has its own kind of stubbornness on a par with the stubbornness of particular material objects.

Primary confrontation in knowledge theory may be defined as the encounter through the present members of a class with that class, and through the class with the absent members of that same class. Put otherwise, it is possible to say that there is a confrontation with both a class and its absent members in consideration of the encounter with present objects.

This is the situation which must be remembered when we are analyzing sense experience. It means that no human individual ever encounters a material object just so and entirely limited to itself. For the act of experience involves the recognition of the class and through the class of its absent members. In the traditional terminology of knowledge theory, the encounter between present particulars (e.g. the observer and his moon) is conditioned by a confrontation involving present universals and absent particulars. Classes are interpolated concretely, and knowing is a way of being confronted. *Universals, on this interpretation, consist in the representation of absent objects.*

In traditional knowledge theory this situation has been given a subjective explanation. It has been said that an element of conception enters into every act of perception. No doubt this is the case; but another interpretation of it is possible, one more compatible with the point of view I have been taking and the kind of analysis I am trying to make. I can say, then, not merely that a conception enters into every act of perception (although it does) but also that *there is a perception of generals.* Universals can be seen, heard, smelled, tasted in the very perception of particulars. This follows from the fact that qualities themselves are general.

A particular material object discloses to the observer its nature as a member of a class when in the very act of recognition of it he is working with the class which he professes to see in it. When the observer looks at a horse and sees it as a horse, he is remembering the similarity between that particular material object and others of the same class that he may have seen or known about on previous occasions and now recalls. So long as in the actual presence of the material object he is aware of the class to which it belongs, he has recognized that the class exists externally, and he does not need to back up his sense perceptions by mentioning the additional and extraneously introduced fact that he is capable also of conceptions.

So far, so good, but the recognition of the external existence of the class has difficulties of its own with which the observer must cope. If there were a single and clear-cut specification of the material object the difficulties might not exist, but the description would then be a very impoverished affair. For the continually interacting and changing world of material objects has dimensions which are not to be encompassed so simply or so easily. No description has ever been given which exhausts the properties of an actual material object. Until we can find some kind of calculus of qualities, philosophy cannot be reduced to mathematics or wholly expressed in the language of mathematcis.

2. Conception

At the end of the last chapter I touched upon a second kind of encounter, which is the encounter with logical classes. In addition to primary confrontation, there is a second and more sophisticated kind. The first kind of confrontation was material confrontation, the second kind may be called logical confrontation, and it consists in moving up one level of abstractness. The observer apprehends the material object as an object but also as a class of such objects, and he recognizes through his apprehension of the class to which the object belongs that there are other classes similar to the class apprehended. He is conceiving now, if you like, instead of merely perceiving, but if so he was led to it by an uninterrupted sequence which started from the level of the sensed material object.

An abstract idea also is an encounter with a material object, for as philosophers of the nominalistic persuasion since Descartes have argued, an idea is always accompanied by an image, and the image is a material object, but it is a material object with a different function; for its function is to represent a class and stand in thought for it. Thus, in a word, the observer is led into considering the domain of classes, which is the domain of logic; and there he cannot help but notice that the relations which exist between classes present their own kind of stubbornness.

The observer encounters material particulars by means of perception, and he encounters logic through his own processes of conception when he studies the relations between classes. But because the imperfect world does not conform absolutely to logic, he thinks it follows that logic is mental. He forgets that he knows logic as he knows the world; through an encounter; and when he remembers that the encounter is also a confrontation, then he ought to be able to pick up the fact that the failure of the world to conform to logic is no evidence that either is mental. Indeed the distinction between his own often fallacious processes of reasoning and the logical relations he is endeavoring to follow is evidence of the objectivity of logic.

The entities and relations disclosed by thought, like the materials and events disclosed by sense perception, fight back when any attempt is made to read them as other than they are. They are equally resistant, equally real. Both logical manipulations and material adventures present the observer with surprises. That is why mathematical discoveries are possible; the sequence of primes, for example. And when something happens that the observer could not have predicted from what he knows, then he can be sure that there is an external world which exists whether he knows it or not, and exists, moreover, in a state of considerable indifference to him. This is what I have referred to much earlier as the inner-induced recognition of subjectivity and it results in the outlines of an objective epistemology which is based largely on a demonstration of outer experience. For now, thanks to the sorting out of the various threads of encounters and confrontations, there is a recognition in what I have called inner experience that there is an outer experience.

3. Three Theses

A number of points can be made about conception by drawing inferences from the foregoing account of the act of experience. I would like first to state them as three graded yet inter-related theses and then to comment on what I think that together they lead to in the way of conclusions.

The three theses are: (a) that what is sensationally immediate may be conceptually obscuring; (b) that what is intermediately vague (distant, yet apprehended through images as partly perceived and partly conceptual) may recede into conceptual generality; and, lastly (c) that what may be too distant for perception in time and space may yet be conceptually material.

(a) The acquisition of knowledge begins with the encounter with a single material object which is discriminated against a background and in this way recognized for what it is. But such an acquisition does not stop there, chiefly because the object encountered was not a valid isolate. Just as the observer comes to the encounter with equipment on his side: with the lessons he has learned from his previous encounters with members of the same class, for instance, and with the knowledge that there are other classes to which that class is itself logically related, so the single material object brings to the encounter all of its close relations with absent objects. It may be discriminated against a background, but this is possible only because the material object encountered differs from the objects in its immediate neighborhood. The object has differences with other present objects but, as we have noted, immense similarities with enormous numbers of absent objects of the same class, objects absent in time on the one hand and in space on the other.

(b) Now when both the material object and the individual encountering it have such complex other relations, it is hardly to be expected that the knowledge which the individual obtains on his side of the encounter would end there. And it does not do so, either, but goes on to comparisons with similar situations in the universe, situations in which there are other material objects, and, it may be, other human individuals encountering them in much the same way. Knowledge is always limited, but the limits recede as knowledge advances.

It is in this way that we are entitled to speak of the Fallacy of the Local Present.

There is nothing merely local about the present, neither in space nor in time, for everything which is peculiarly related to it participates in it even in the most primitive aspect of sense perception. Thus the kind of happening we have come to call 'having a sense experience' is the beginning of an accumulation of knowledge which always exceeds the immediate occasion.

Another point might be made somewhat as follows. Considering the dependence of the observer on his environment — on his physical environment for air and shelter, on his biological environment for food, and on his social environment for beliefs and language — it could fairly be said that he himself consists in an arbitrary selected segment of an environing natural world. There are no valid isolates, each is part of all; and so all of the effects of the known material objects acting upon that other material object which is the knower must be counted as fundamental. Consciousness is from this point of view the center of the impingement of the world upon the observer, who is an uncommonly sensitive mechanism for sorting out tropisms and integrating influences.

The observer is both a material object and a knower with a view of himself and of the world with which, as a set of material objects, he interacts. So much is plain; but there is often something more. For subjectivity can also mean the condition of being in a position to observe that object itself from the inside and some limited portion of the environing world from the outside, both under the dictates of a judgement which renders the view of the object from inside more real or more general than the view of the environing world.

In what may be called a second subjectivity, which is where most of us take up our position, consciousness means the observation by the subject of that subject interacting with the objects in the world around him (1). Often it is possible to interfere and to change the interaction somewhat, but just as often it is not; and in the latter case the observer is a passive onlooker at the sequence of events which may or may not be to his liking.

(c) It happens sometimes that this judgment is wrong. The self is a point and a focus, nothing more; not anything in and of itself. For there are no objects in the mind, neither material objects nor logical objects, only the knowledge of objects constructed in terms of images and ideas. The objects of which minds have knowledge must be independent of the minds which know them for there to be knowledge at all. Mind is simply matter in a condition to be aware of itself and of the world beyond, including the representations of absent objects.

The observer's organism was constructed out of elements selected from the world and put together in such a way as to accept the world. It is his perspective — not only the angle from which it is taken but what lies within it — that gives him back to himself as a self. For the self is not to be found through internal concentration but rather by

1 See above, chapter VIII, section 2.

deducing from the world the peculiar perspective from which it is being observed, that part of the world which can be apprehended from the self. The self itself is only a point-instant from which it was possible to make a world selection.

4. Assisted Conception

In the previous chapter, in which I discussed perception, I called attention to the extraordinary service which artifacts have performed in disclosing the existence of the three-segmented universe. The external world as ordinarily perceived by the unaided senses is apparently extended in both directions, linked in an uninterrupted fashion to the smaller elements and processes of the microcosm, and to the larger elements and processes of the macrocosm, the knowledge of which was made possible by instruments of observation designed for the purpose. Now I must indicate a corresponding development in which conception is similarly aided.

Just as observations conducted by perception were extended by instruments, so thoughts in the process of conception have theirs. That a man is able to think about the world at all is due to some transformed segment of it. Some instruments have been around for a long time: books, for instance, while others such as computers and their special languages, are comparatively new. Thought processes aided by mathematics to a large extent may be characterized at this level of social involvement as one of interaction with the artifacts of conception.

For thought processes still take place in relative isolation as the work of single individuals deliberately concetrating on them, but to imagine a professional engaged in solving one of the problems arising in his field of endeavor as a single, solitary and private affair would be to make a wholly misleading analysis. On the assumption that everything in the mind was first in the external world, except error, an individual thinking is that individual deliberately closing the door to further impressions and workings inside the mind with what had been brought there previously.

There is some material and even social content ot even the most absolute of solipsists. Husserl's 'transcendental-phenomenological reduction', in which he traced the ultimate elements of experience to certain intentional performances of the subjective ego by the method of inner observation, overlooked the original external source of the individual things whose general characteristics, such as sounds and colors, he noted. The abstract thinker excludes all further influences from the external world while he is thinking, but this does not deny the source of contents of his thoughts. It simply means that the process of 'bracketing' was an attempt to exclude further influences.

The conceptions of ordinary experience work with only the artifacts of language; most thought, if not all, is made possible by language. But the influence of artifacts has been extended to material tools; books, after all, are only collections of recorded language. The same cannot be said for computers. Machines who think, as they have been called, raise for the first time the whole question of technologically-assisted conception.

Man fashions artifacts in aid of every one of his endeavors. Conceiving is always an effort to understand some segment of the material world. Since the brain is also composed of material, what is involved at some point is a concentration of the laws of the universe in it in the form of knowledge.

5. The Mind and Its Objects

In concluding the arguments offered in the previous chapter and this one, I want to offer a few summatory remarks. I found it necessary to make a distinction between the mind and its contents, just as I had to do for consciousness and its contents in the previous chapter. A typical mental state consists in being aware of an object, regardless of whether that object is internal, as is the case with an idea, or external, as with a chair. The contents of a mental state is what Meinong called the *vorstellung*, what stands before it. As I noted earlier, the mind is made possible by the brain, and the brain is matter in a highly complex condition, how complex we are just beginning to understand. The computer model of the brain has been proved insufficient by the discovery that there are some twenty peptides which act as neurotransmitters (1).

It is possible of course to have a brain without a mind, as happens with a patient in a coma, but it is not possible to have a mind without a brain. Since the brain is a material object, I shall have to describe the mind as *para-material* in some as yet unspecified way. All subjective interpretations of mind are out of date: there is no independent 'mind'. The evidence for this is now overwhelming, and comes from many sources, such as neuro-surgery and mood-modifying drugs.

The only secure domain of the mind so far as reliable knowledge is concerned is what can be abstracted from the world of matter and energy in space and time, the material world. The material and the physical, it may be parenthetically remarked, are often confused. Matter is the substance of all actual organizations, while the physical is the lowest level of organisation. As noted in chapter II, other levels are, in ascending order, the chemical, the biological, the psychological and the cultural. The physical, then, describes only the lowest of the material levels.

Material objects at all these integrative levels constitute the working field for the acquisition of knowledge by the mind. I mean by 'material objects' not only those which exist now but also those which have existed in the past and those which will exist in the future, including those made by man as well as those found in the natural state. This is the inventory, the pool from which knowledge is abstracted. It includes, though in a more unspecified fashion, what may exist as well as what could or will exist counting all the possible worlds which have been added to our conception of the actual

1   J. Hughes, (ed.), *Centrally Acting Peptides* (New York 1978, University Park Press).

world. The two together just about sum it up, though of course that leaves us with many problems and puzzlements.

Having 'knowledge' means after all holding a representation of some portion of the environment: its details and their similarities. It is highly selected, limited and approximate. Knowledge is ultimately based on experience (including not only crude sense experience but also knowledge from diverse other sources, such as instruments, books and other individuals), and it is corrected continually.

The individual has a way of dealing with those possible objects which are not present, with absent objects. The status of absent objects as part of what the mind knows is a topic which I have already discussed, but there is another class of possible objects selected from both present and absent objects. We may for want of a better name cell them 'impossible objects'. Meinong treated them at length, and gave as examples 'golden mountains' and 'round squares'.

What is allegedly involved is of course a confusion, for a contradiction cannot be charged to any material thing. Somewhere as yet undiscovered there may be a mountain made of gold; it is not inherently contradictory; but a round square, of course, is. While certainly Meinong's 'golden mountain' and 'round square' do not exist, gold and mountains, round things and square things, do. We must have a special category, then for what could not exist apart from what could or does, and our resident domain for the recognition of it is the mind.

Mental objects are not all of a kind but all are reflections of artifacts, and they do offer some variety of existence. It is not easy to form an image of a round square, if indeed it is possible to do so at all; but it is possible, and as a matter of fact quite easy, to form an image of a golden mountain. The mental state is a private state although thoughts can be communicated, and the contents of thoughts are not always kept private. The feelings are private, though we know by inference and analogy that something like them occurs in other individuals.

6. Coping with Knowledge

What is called 'the mind', then, is not a single function but a loose collection of functions to which we have given a common name. What is meant by the term is that sense impressions are received, retained in the unconscious, and combined in thought. They are stored as beliefs and revived as memories. They are even projected onto the world, from which they came as raw impressions, by means of the will and by impulses to actions. 'The mind' is only a description of the way in which we examine a segment of the world in abstraction from the experience by means of which we obtained the knowledge of it. We have the capacity for thinking about what we have learned and so of learning more about it.

When I say that the individual copes with knowledge, what I mean is the way in which

his brain copes with information, how it reacts to input. The object of coping with knowledge is the survival of the individual, immediate survival first and then ultimate survival. Immediate survival is more *importunate*, ultimate survival more *important*. With respect to knowledge, the human organism is a self-regulating mechanism based on the workings of subordinate systems. The reception, organization and application of knowledge is one such system, but there are others, such as the temperature of the body, the regulation of the rate of blood flow, etc. Some of these systems are autonomic, some are not: the knowledge system is not.

'Mind' may be described here as 'conduct toward knowledge in a living organism', more specifically, as the way in which man deals with knowledge. This includes its acquisition (consciousness or alertness), its storage (memory), its manipulation (thought) and its use in the initiation and direction of action (decision or will). In perception we receive knowledge, in memory we retain it, in thought we conceptualize it, in behavior we employ it. It operates only in the presence of an immediate environment of sufficient variety to provide the necessary novelty required by alertness.

Mind, in other words, is the general name for that process by which the human animal moves from sensations to universals, from the barest of experiences to the possession of general knowledge, and from thence to the employment of force. Thus every mental function consists in dealing in some way with knowledge. It is not possible to think of the mind apart from it.

I have already defined substance as 'the irrational ground of individual reaction', matter as 'static substance', and energy as 'dynamic substance'. Mind, then, is energy at the psychological level of organization. The distinction between mind and matter should now be clear. Every material organization exists at some energy level.

Expanding an earlier conception, I might add that there is nothing in the mind that was not first in the external world except the capacity to acquire, assimilate and deploy knowledge, and to devise error; a capacity, in other words, for dealing with truth and for inventing falsehood. The mind has two capacities, it can apprehend truth or create error. Strictly speaking, error is not 'created' but invented, and consists in a false selection and arrangement of genuine elements.

I would include in these capacities the ability to organize truths by combining or separating them. Also I am speaking of course of content. The ability to apprehend truth is mental. Thought is mental, and there are mental states, such as belief and doubt. What is in the mind in the way of truth is of course both partial and general. What is lacks through partiality it makes up somewhat by being general.

By definition the mind is incapable of comprehending the powers which it has as a whole. For consider: if it comprehends them as a whole then it extends beyond the whole, and if it extends beyond the whole then the 'whole' is not a whole. Thus the mind is incapable of apprehending itself as a whole, and so by definition can never understand all of its own operation, though this must be read only as a limiting case.

Clearly, then, the question of whether there could be an accurate model of mental operations is left somewhat ambiguous.

We rarely attend to our mental states but, on the contrary, it is the mental states by means of which we attend. Even thoughts, which command our exclusive attention and concentration, are not entirely mental. We think *about* something, after all, and what we think about can be traced to one of the two tiers of the two-tiered external world: the ground level of individual material objects or the upper story of logical objects (the entities of logic and mathematics).

When I say, then, that there is nothing in the mind that was not first in the external world except the capacity to acquire, assimilate and deploy knowledge, and to devise error, I am entirely aware of the large inventory that this implies. The external world we refer to so often in knowledge theory is another name for nature, and nature contains more than any of our limited schemes. It is not only the case that men are ingredients of nature but also that they emerged from the background of nature as themselves natural objects, albeit objects of a special sort. Thus everything that enters the mind of man is natural.

Man himself could almost be defined as the animal that invents falsehoods and then reacts to them in terms of essential truths: hence the arts for example. Accepting truths is something he does in common with other animals even though his truths tend to be much more abstract. The difficulty contained in the subjective view entertained by the idealists is probably due to the fact that we have attributed to our powers of invention much that was due instead to our powers of discovery. It takes great skill to find and then accurately to describe conditions which are both contrary to fact and crucial.

The mind of the individual is subject to two sets of influences. One set issues from his genetic history and the consequent adjustments of the biological organism, the other is epigenetic and issues from the immediate environment of culture; one is internal, the other external. Both are organized by the mind. Sanity depends upon the ability of the mind to make a single synthesis out of its various influences.

The structure of the organism is generally credited to its genetic origins, and no doubt there is something to the claim. But the point is that the organism is whatever it is regardless of how it got to be what it is. In any cross section of time we have the spectacle of a mind receiving stimuli from the material culture as well as from within the organism. Those from within are apt to be chiefly organ-specific, whether it is a feeling of thirst coming from dry tissues or pain from a sprained ankle. In both connections the mind functions passively, as a mere receptor of stimuli. Only after that is there either an instantaneous or a delayed response.

The current attempts of the structuralists men like Chomsky, but also Barthes, Levi-Strauss, Foucault and Lacan, to understand the brain by attributing to it an inherent grammar rests on nothing stronger than the recognition that the members fo the human species employ languages in most if not all of their operations. The dissection of the brain is a very primitive kind of analysis at the present time; and though no doubt progress will be made in this direction, it will be a very long time before the presence of syntactical structures are discovered in the neurons, if indeed they are there.

Anyway, it is a serious mistake to confuse methods of communications with systems of ideas. Complex mathematical systems were discovered, not invented, and they were not discovered for the sole purpose of communicating them. Logic is no more a part of the brain than the material world is a part of our senses, though the brain is instrumental in discovering logic and our senses are instrumental in discovering the material world.

CHAPTER X

MORALITY: THE GOOD

1. The Fact of Moral Practice

In chapters VI and VII we looked at man as he is when determined by both internal and external forces, that is to say, by his own requirements and by those restraints which are imposed on him by his material and social environment. Now we must look at him again, but this time considering him a free agent. What are his moral options? These in my opinion are not static affairs; they are the result of his use of freedom in the course of his involvement in practice.

We should have learned by now from the social studies that outside the physical sciences there has been very little understanding of the scientific method. It may well be that its application to a social subject-matter involves too many variables. There is no isolation of the kind experiment requires and there are as yet no viable abstractions. In any case, nobody has succeeded very well in finding invariantive social laws.

Philosophy as a social undertaking has fared no better. Admittedly, a great deal has been written in philosophy about the scientific method, but, as I noted in the first chapter, most of it has been in the direction of its interpretation in philosophical terms. Few thinkers have ever turned it around to see of what use the method of science could be in philosophy.

There is a valid reason for this. Most studies in philosophy have concentrated on its abstract formulations. Except in terms of politics, few between the ancient Greeks and the nineteenth century Marxists and pragmatists ever thought of a connection between philosophy and practice; such a conception was foreign to the philosophical tradition, which does not usually contemplate the practice of philosophy. If the question of morality was introduced into social life at all, it was in the name of some other institution, usually that of religion. There were exceptions of course, but speaking generally, philosophy, it has been held, was something one thought *about*, not something one *did*.

Any attempt by the philosophers to look at their subject-matter the way that the physical scientists have looked at theirs would have to begin with the empirical observations of how philosophies have been employed by other disciplines; that is to say, how assumptions about the nature of reality have been incorporated in the constitutions of governments and the creeds of religions, revealed by what tools and linguistic expressions societies have chosen, by what they have taken for granted in customs and traditions.

The individuals in a society often face a more complicated problem, for in addition to the assumptions of the society, which presumably they share, they often have a number of private convictions. Thus the fundamental beliefs of the individual as well as the ideas revealed by his behavior patterns may be divided between those he holds in common with other members of his society and those which are his alone.

What is true of theories of reality must be true also of ethics. At the empirical level of morality it should be possible to deal with motivation, with behavior, and even with artifacts, in a way which has not been adequately done. Morality moves now in an atmosphere of empiricism and in the midst of an environment of artifacts. As life in civilization becomes more complicated, so must morality, at least if it is to keep pace with these new developments.

There are no direct interventions by means of which we can say, this is how science translates into morality, none at any rate of the clear kind it is possible to find in metaphysics; and so the impact of science on morality has to be deduced more subtly, perhaps, by approaching it from the assumptions underlying the physical sciences and technology. That is what I have endeavored to suggest here. An empiricist, of the school which recognizes the importance of laws in physics, is the model I have chosen, and his presuppositions are the ones I have taken for granted.

2. The Moral Integrative Levels

Ethics broadly speaking is the theory of the good and the bad. I define the *good* as what is needed, and the need can be for any existing thing. For every need there is a drive to reduce it. Thus the good is quality external to anything to which it is a good, that quality which is a bond between wholes. It makes a contrast with the beautiful, which is an internal quality, that quality which exists as a bond between parts or between parts and whole. In other words, ethics studies the quality of completeness while aesthetics studies the quality of consistency. But both the good and the beautiful are internal to the cosmos; which entitles us to say that there is only one world-quality, for while beauty is more intense, goodness is more pervasive. The cosmos as a whole can be beautiful but cannot be said to be good because with our severe limitations we know of nothing that it needs.

We shall be more concerned here of course with what is good for individual man, his needs and his drives, which is the topic of traditional ethics, and I will touch on the others only in order to put the individual's good in its proper perspective.

The good for man is what is needed by man. Obviously 'need' is a two-way relationship and so involves privileges as well as obligations: what man needs to take from other things and what they need from him. How he deals with his encounters is what makes up his moral life. The legal names for privileges and obligations are 'rights' and 'duties'.

Privileges and obligations are graded in terms of confrontations. For the individual there are four levels of such confrontations, named collectively the moral integrative levels. These are:

the cosmic,
the human,
the cultural, and
the individual.

Each of the moral integrative levels has its own relations with the other three. I won't try to cover all of this involvement but will confine my exposition to the individual. I will deal with his relations only, because it is the ground level upon which all of the others depend. The orientation adopted here will be that of the encounters of the human individual with himself: what he needs from, and owes to, himself, and — reading up — from his society, from his species, and finally from the world.

3. The Individual's Encounter With Himself

The first level of confrontation of the individual is with himself. The major decision respecting existence was made for him by his somatic organism: on its own it strives to live. 'Is life worth living?' someone asked in Samuel Butler's day, and he replied in his *Notebook* 'This is a question for an embryo, not for a man'.

There is an authenticity to the domain which defines what the individual can expect of himself and what he owes to himself, beginning with his duty to preserve himself and stopping short only at his right to a share in his society. Since not all individuals are alike, the individual in preserving himself is saving something of value which is unique. Thus he must first reckon with himself as the primary fact of his existence. His direct dealings with himself are basic but minimal; they consist cheifly in receiving sensations, for what else could he manage without mediation? He could think or perhaps talk to himself, but that requires the artifacts of a language. He could perhaps scratch himself unaided, but only in front; in the back he would have to have the assistance of a back-scratcher.

Private feelings unrelated to anything outside his body are at the lowest sub-level. Before anything else can be done he must attend to himself, he must cater to his own needs. The motivation of all human behavior is best understood as the effort of the individual to complete himself by satisfying such needs. To say next that the good is what the individual needs is not to equate it with the need itself nor even with the relation between the individual and what he needs. It is the material thing itself in so far as it *is* needed.

The good as what is needed cannot always be a matter of mere pleasure or even happiness, but more often something is required in order to complete existence, the result of organ-deprivation. The individual acts as an agent for the needs of his organs. The reduction of organ-specific needs benefits the individual's whole organism. Every need activates a drive, and it is the drive we recognize as behavior, for it always involves the use of power of some kind, everything in fact from physical force to psychological persuasion. The use of force may be good or bad, but there is no good without it. Since most of the things needed are contained in the environment, to reduce all of his needs an individual would have to dominate his environment -all of it. Thus his most overall exigent need is for domination, and its drive is expressed through aggression, which can be destructive as well as constructive.

The large and continuous dependence of organisms upon an interchange with their environment is no less effective whether the need be for air, food, water, clothing and shelter, in prolonged and repeated supply, or for more complex articles to serve more intricate needs, such as knowledge, activity and security. The minimal good is of course asymmetrical: X is good for Y, and most goods are like that. Any complete good would have to be symmetrical: X is good for Y and Y is for X. Individuals need friendship as well as food, but the first need is symmetrical the second is not. Friendship between two individuals may be good for both of them; but while catnip is good for the cat, the cat is not good for the catnip, which is destroyed.

Technology enters ethics as a subdivision. Drive-reduction almost always involves artifacts: not many activities can take place without tools or signs (or both). Thus what is to be made and how it is to be used are primarily moral questions. and moral considerations are very much involved in the design of artifacts, their production and operation. What men do is seldom done with their bare hands, and so artifacts are involved in nearly all human endeavor.

In moral terms what is a human individual and how does he go about his drives for need-reduction? Though usually considered for most purposes a highly integrated whole, he is, as we have noted, a loose organization, a democracy of sorts, consisting in a compromise of voices and persons, in which the critical self settles with the impulsive self for what it must accept of what is available; that is to say, of what is offered or of what the individual operating under his own power is capable of appropriating.

In the course of reducing his needs the individual will have many encounters with need-reducing objects. More than one will press for attention at the same time. On each such occasion he will have to choose. This then is his first level of confrontation. Morality is not concerned with activity in all of its aspects, only with behavior so far as it involves confrontation. As that term is emplyed here, it means the outcome of any encounter conditioned by the specific stipulations of type responsibility, and involves the good.

There is of course always more to it than that. Confrontation represents the fact that the moral act is not a valid isolate; the entire involvement of existence is always represented through the operation of some agency.

At the ground level, then, individual confrontation may be defined as *conduct toward another, in consideration of all others*.

The first confrontation of the individual is with other members of his type. Membership in the same type involves a certain responsibility which will be discussed again at each of the other three moral integrative levels. The relevant principle may be stated as follows.

All individuals of a given type are attracted to (and hence involved with) all other individuals of the same type in virtue of a common dependence upon type responsibility.

The individual in virtue of his mere existence incurs a moral obligation to maintain himself with integrity and dignity. A 'good individual' here means one who is good for himself. This is in effect the individual level of morality faced inward. The individual is a whole to his parts. His obligations at this level are those which he owes to his parts and through them to himself. Among his parts must be counted his organs with their specific needs and drives.

Now since the individual can do only one thing at a time, he can engage in the drive to reduce only one need at a time, while the other needs are suppressed temporarily. Thus there must be competition as well as cooperation among the needs and a procedural order arranged for their respective drives. Does food come first, or a mate? Is the search for knowledge to be subordinated to ambition? Harmony among the needs and drives is also an obligation, one the individual owes to himself, without which there can be little hope for the good life.

The establishment of harmony among his parts is the first obligation of the individual. It results in opportunities to engage in privileges. By means of self-knowledge of this sort, a large and powerful measure of self-control is made possible. And when the individual controls himself he is in a position to act integrally and with greater force.

Speaking only about the ground level, the hedonistic morality is the one proper to the individual. To be a hedonist is to lead a life of pleasure and there is pleasure to be had from *all* of the organs; that means from the brain and the musculature as well as from the sex organs. Knowing can provide its own pleasures just as much as eating can; while the joy of physical exertion is quite common.

It is here perhaps that privilege and obligation merge. The organs exist in order to be used, that is a privilege; but the individual has an obligation to himself to use them all and not merely some in preference to the others. There are conditions imposed on hedonism, an ordering of gratifications which would not be the same for all individuals. The generalized morality which must result is as far as hedonism can go, for it is still focused exclusively on the individual's own needs, on *his* goals.

The individual does not deal with himself as a kit of parts. 'Self-realization' is the name F.H. Bradley gave to the moral attitude and behavior of the individual toward himself as a whole person (1). It involves self-protection and self-completion: the individual's duty to himself is to perfect and complete himself in some way which shall be altogether consistent. He aims at himself as a whole. This is the authentic level of the individual even though it is one not capable of being made absolute.

To be able to regard himself as a person means to the individual to be a whole containing parts, to be an entire man. The aim at entire man is where morality begins, but it is very far from being the whole story of the moral life, for there are other encounters. To accomplish most of his need-reductions the individual must go outside his own body: what he needs must be obtained from some portion of the world. He is in any case engaged in a constant stage of interchange with it and is in no way independent of it. He is a part of it and must look to it, and this requires of him some degree of what Nietzsche called 'self-overcoming' if he is to function as a member of society. Thus the good for the individual must include his service to society.

4. The Individual's Encounter With Culture

The next stage of the discussion therefore must be about social ethics, and for this we need a definition of society. I define a society as *that social organization within a culture whose boundaries are recognized*, chiefly by a common language. It would follow from this definition that while 'society' is not a valid isolate, 'culture' is. Thus 'France' is the name of a society within the European culture. The individual's primary aim is the service of society, and he does so in most cases by contributing to its culture. His secondary aim is to derive such satisfactions for himself as are not inconsistent with the primary aim.

As we have noted from considering the moral integrative levels, there are vistas beyond the cultural, such as the species and the cosmic; but the actual practical life of the individual is lived for the most part as a member of some one particular culture, and it is in that culture that all of his struggles take place.

1 *Ethical Studies*, Essay II.

There are ingredient in every individual two moral codes, his own and his culture's. The first is the result of his private experiences, the second of his cultural encounters: those with friends, tools, customs and institutions. His practices are the result of the skillful blending of the two codes. In a smoothly functioning culture there are few if any conflicts; what he wants to do would be also what the society expected from him or at the very least allowed to him.

Social morality may be defined intrinsically as the quality of the internal relations of a culture. The social good is the dominant inner quality which emerges from the integrity of organization of that society. It has a name, one used infrequently: the 'ethos', though many a stranger to a culture has noted its peculiar atmosphere, its pervasive affective influence.

It should be possible now to state the principle of type responsibility as it applies to the individual in his relation to his society.

All individuals are attracted to all other individuals in a given culture in virtue of a common dependence upon that culture. This is type responsibility at the cultural level.

Confrontation at the social level does not require of the individual that he considers anything beyond but compels him to take into consideration all other members when dealing with any one. Thus we have his *conduct toward another individual, in consideration of the whole culture.*

A culture always has two aims. Its primary aim is to contribute to the whole of which in turn *it* is a part, in a word to humanity. Its secondary aim is to enable its individual members to assist it in its primary aim. Thus its secondary aim is to contrbute to the welfare of all its individuals. This it does through the establishment and enforcement of rules of order in so far as that order and those rules are consistent with the maximum possible amount of individual freedom. By behaving morally toward other individuals the individual himself helps to construct a culture. The social good is inherent in the advantages which the culture bestows upon its individual members.

Human individuals constitute the first ingredients of a society, artifacts the second. Individuals within a society agree to use the same kind of technology; indeed the type of communication defines the social boundaries. Communication is chiefly by language but may also be by other kinds of tools. Individuals rarely deal with each other directly, it is more usual for them to deal through artifacts. Society is composed of institutions, and its individual members usually belong to them and use the tools appropriate to them, for institutions are the principal owners of artifacts.

5. Societies as Repositories of Artifacts

The influence of artifacts in human life is nowhere more in evidence than it is in social organizations, in society as a whole and in politics. The Marxist revolution in some Asian countries, notably first in Russia, was occasioned by the discovery of the factory system of economic production. The change came in the least industrialized countries and in the name of serfs rather than of industrial workers, yet the doctrine was suggested by the new industrialism, the advent of the technology of a machine age in social development.

Social organizations, small or large — and of late the tendency is to have large ones commensurate with the enormous sudden growth in populations — always own or control artifacts appropriate to their size. The state includes not only the judicial system and the army but also the instruments upon which each depends. Taxation puts at the disposal of the state large sums earned by the citizens through their productive activities, and such sums represent material power: the ability to control, purchase or distribute artifacts. The more powerful the state the greater its ownership of artifacts, regardless of whether it uses them for or against the welfare of its citizens. Thus to a large extent it may be argued that technology and its products determine the shape of politics and of social events. The customary name for artifacts in society is 'property', which may be defined as material goods, with ownership understood as the exclusive right to their use. This holds true whatever the economic system: private enterprise or socialism. 'Property', in short, is the name for artifacts considered in their social role. Along with it go laws, sentiments, observances; there is always a good and a bad way to use property. It is a mistake therefore to think of society as a mere collection of individuals; without artifacts they would have no way of getting together, and so we must think in terms of cultures instead. Morality, understood as correct use, is always established in some way; it is fixed, and so furnishes stability to society. It underlies the principle of order and so is what makes of a society an autonomous community.

The Marxists and their followers have understood very well the necessary association of society with its artifacts. Unfortunately, they have understood this relationshop almost entirely as a struggle for control conducted by economically diverse and unequal classes. The artifacts, they have insisted, must be publicly owned and class dominated for an unstated interim. The briefly mentioned goal of doing away with all classes, and with them the state, is put off into an indefinite future. Industry and its development of large-scale artifacts is what precipitated the crisis in social organization and threatens to compel a change everywhere. Thus far no one has proposed an acceptable alternative, only used the large-scale artifacts of industry to justify the most rigid and ruthless of dicatatorships and the promotion of violent revolutions.

Not that anyone can deny the importance of artifacts in the immediate environment; they are crucial to all human associations but they are not exhaustive. From a society's point of view, the total environment may be divided into an available environment, which is that portion of the total environment with which it is possible for a society to interact; and the immediate environment, which is that portion of the available environment with which the society does interact. All this is made possible of course by the tools and instruments which the interactions generate; these make social life possible and are therefore essential to it. Such material goods are needed by a society, for, as Aristotle observed, 'it is impossible, or not easy, to do noble acts without the proper equipment' (1).

Morality at the social level is in effect the workability of a society. There is only a short logical distance from the quality of the good to the relatedness of the useful. We have noted already that the justification for morality lies in the fact that the conflicting drives between individuals who are striving to reduce their needs compel some degree of cooperation. Morality exists in order to facilitate the pursuit of goods by regulating individual competition.

In pursuing need-reduction an individual encounters the guidelines of his society. Morality for the individual living in a society is a matter of having his preferences governed by principles. When his interests are aroused he acts, and his actions are regulated. In the course of a moral life, he finds that there are rights which he may conventionally expect to receive from society, and duties which he may be expected to give to it.

Moral rights are laid down by a society; what is right is what the society approves. The morality is designed to facilitate the need-reductions of the individual by providing ready access to the requisite materials. But he also has duties, which are obligations to act in conformity with the established moral order as represented by law. Such conduct collectively makes the very existence of the society possible.

The laws of a society represent its attempts to codify its morality, for, again quoting Aristotle, 'it is through laws that we can become good' (2). It is the task of the legal establishment both to codify and to enforce conformity with the prevailing social morality. Law differs from morality in two ways: in the first place it is deliberate and known; and in the second place it changes slowly. The morality of a society is an informal affair and enters into all of its structures and activities.

What may be called 'the implicit dominant morality' of a culture is to be found in at least five places: in the conscience of individual members, in the hierarchy of institutions, in prevailing sets of preferences, in the pervasive qualitative atmosphere of the ethos, and finally in the formally established laws. 'Conscience' is the reference to the implicit dominant morality as it exists in the subconscious. The hierarchy of institutions in a society is its response to the implicit theory of reality which ordered them. The prevailing sets of preferences are qualitative matters of taste which were implicitly acknowledged and covertly established. The atmosphere of the ethos is how the implicit dominant morality is felt by anyone entering the society. And, finally, the enacted laws of a society provide the formalized version of its morality, which explains why those laws rather than others.

1   *Nico. Eth.*, 1099a31. See also 1101a15.
2   *Nico. Eth.*, 1180b25.

6. The Individual's Encounter With Humanity

The individual encounters himself first and then his culture, but the encounters do not stop there. Beyond any limited culture lies humanity, of which every culture in turn is a part, and beyond humanity there lies the vast universe. Any consideration of morality must deal also with these. I turn next, then, to a consideration of the moral integrative level of the human species.

Humanity is the name for the sum total not only of human individuals, but also of social groups, artifacts, institutions, nations and civilizations. Each moral integrative level has its autonomy, and so there are many other considerations which cannot be overlooked in the consideration of humanity. To think that they can is the flaw inherent in 'humanism'. The individual belongs *sui generis* to humanity merely in virtue of his being human. This is something he *is*, an inescapable condition of his being, and not merely something he *does*. Yet cultural relativity rests on the solid base of cultural conditioning, a fact which must affect every individual's existence.

What is peculiar about man and marks him off from all other organic species is his *recognition* of his type responsibility. The universal character of confrontation for the human species involves the implied representation of absent members. Most are absent in time as well as in space, but they are present in what might be called the individual's surrogate encounters with other members.

In describing the lesser encounters the individual has with himself as a person and with his culture, I pointed out that the common dependence upon type responsibility occurred as 'conduct toward another in consideration of all others'. At the level of the whole of humanity this principle becomes one which can be stated as follows.

An individual exists among other individuals in virtue of those of his fellows who are not present and for whom through the species he is responsible.

In a way, humanity interposes itself between the individual and his culture. The individual, in other words, never acts entirely for himself because his individuality is not independent, not absolute. He is not only a citizen in a particular society but also a member of a species, an example of a type.

Confrontation theory at the level of humanity may be formulated as: *Conduct towards culture, in consideration of all other cultures*. There is thus a division in the self into at least three parts: a short-range self, or his involvement with his culture, an intermediate-range self, or his involvement with humanity; and, as we shall see, a long-range self, or his involvement with the material universe, the cosmos.

Morality exists for the human species through a total commitment to the universal. All individual and social differences, all artifacts, institutions and societies, lose their particularity at this moral level and merge into the general. Individual men are replaced by, and subordinated to, Man.

What human individuals have in common, then, is their sheer humanity. Not the way in which they live nor the customs they follow, not even the beliefs they hold or the pleasures they prefer — all these are as nothing when measured against the weight of the basic facts of their existence, which is that they are born, grow and develop, decline and die, against a background of intermittent pain. What people have in common, whether they know and like it or not, counts for more than their differences. In the end it is man who matters most to men. There is some consolation in the recognition of a common brotherhood of suffering. It is pain, and most of all the final pain of death, which makes the whole world kin.

A civilization is a culture writ large. There is as yet no global culture which would make civilization equivalent to humanity, at least in the limited sense of terrestrial existence. But in any case, since humanity covers all those individuals, cultures and civilizations which have existed as well as those which will, the term still retains its generic meaning, and we have to deal morally with its encounters with other organic species, just as we discussed earlier the individual's encounter with other individuals and the culture's encounter with other cultures.

All living organisms share a bond in virtue of their common possession of the property of life. Thus for any given species there is a moral obligation which it owes to all other species. The recognition and fulfillment of such an obligation is what I call *species-responsible behavior*. The situation is not a simple one, however, for the higher species live on the lower, and man on all of them. Without sufficient aggression he could not kill and eat the members of lower species, but neither could he survive unless he did. The food chain is a fact of life for all organisms, humanity included; so that to some extend diet defeats all moral theory. The contact of other species with humanity does not seem to do most of them any good.

As man looks downward in the organic species for support, so he might look upward to achievement. As the member of a species his obligation is to develop a higher species, as much higher than man, say, as man is higher than *Australopithecus* or the earlier apes. The direction of the organic species, looking back upon origins, seems to lie in the increase in capacities and potentials. If man strives to perfect himself, he may, through the biological sporting of genetic mutation and given a sufficient number of incidences, in time produce a species superior to himself. That is how the progress occurred in which man emerged and how he might produce a higher species.

Humanity is an organization within the greater organization of the cosmos, with privileges from it but also with obligations to it, obligations which argue against the absolute autonomy of humanity. The workability of that rough organization which is the human species implies an ethical ideal but also points to a condition of imcompleteness in any local and immediate sense. In a word, no limited society can be equated with humanity, though humanity until now held itself to be unique in the cosmos. The current estimates that there is life on many other planets in remote galaxies scattered through-the material universe does make havoc of the elaborate pretensions to exclusivity which have been hitherto the custom of life on earth.

7. The Individual's Encounter With The Cosmos

In general, the good for humanity is what is needed by humans, but what is that exactly? What does mankind need in order to survive? Certainly at least a mild climate, with a temperature ranging somewhere between 0° and 100° F, an atmosphere containing the right amounts of oxygen, nitrogen and hydrogen, a soil suitable for the support of living organisms, and a self-sustaining cycle of fauna and flora.

Humanity confronts its environment, physical, chemical and biological, with the good as it appears at the interface where mankind meets the non-human. In this way the two segements of nature show a natural cleavage. Humanity in this perspective proves to be a small part of a large and all-encompassing natural world in which all parts are bonded together by structures developing their emergent qualities. Morality becomes largely a matter of relating to the nearby environment. But the earth 'needs' the solar system and the solar system 'needs' The Galaxy, and, in all likelihood, The Galaxy 'needs' the astronomers call the 'local group' of galaxies, etc.

The moral integrative levels are constructed out of the recognition of the many confrontations of the human individual. An individual has to live with himself, he has to be a part of some culture, he has to face up to his membership in the human species, and finally he has to exist somewhere in the cosmos.

Despite the immensity of the universe and the puny nature of the human organism, the fact remains that the individual's 'need' of the universe is as organ-specific as his other organic needs. His need for ultimate security is a need of his whole organism, and it calls for a kind of super-identification with the universe or its cause as the only possible objects which could offer opportunities of permanence. Because he wants so eagerly to aggrandize his ego, he appeals to art to transcend its social context by reaching for cosmic values and appeals to science to learn about them. And because he wants so strenuously to survive, man would identify with god; and this may well be the source of all his religious strivings.

The individual when confronted by the cosmos responds with the construction of the fine arts and the pure sciences. By means of the fine arts he seeks to bring into culture representations of cosmic value. All of the great works of art are in this sense religious symbols regardless of whether they were intended for the service of some particular church or not. And by means of the pure sciences he strives to discover more information about the cosmos. Astronomy comes to mind as perhaps the most vivid example. It is a moral gain also to increase the amount of beauty in the world and to add to knowledge.

The good is the bond between material things. This time I use the term 'bond' in order to bring out the aspect of quality, for the good is above all a quality. Values are qualities high in the scale of moral integrative levels. The beautiful, for example, is the bond between parts, so that every whole which is good possesses a degree of beauty depending upon how well its parts are integrated in the whole. All wholes are good for other wholes and all to some extent are beautiful in themselves. Thus goodness and beauty are coordinates of all wholes except one: the cosmos is beautiful; whether it is also good requires a knowledge of what lies beyond it, and that we do not possess. We know only that it is beautiful and that anything less is good.

Does it not come to this, then, that every quality is at once both good and beautiful? The good consists in cosmic connections and the beautiful in how well these cosmic connections are organized. Assuming that the cosmos is unlimited, there could be nothing outside it for which it could be good. Kant said he was impressed by the starry heavens above and the moral law within, but he failed to see that there was any connection between them except such as he could derive from the consciousness of his own existence and the unknowability of the thing-in-itself. He could find a connection only by assuming that the moral law was independent of all animality even though he knew himself to be an animal dependent upon the planet earth. He was blinded by his subjectivity to the connections between individual man and the cosmos in the very terms he explicitly rejected: the continuity of material being with its multiplicity of forms.

Given that every material thing in the universe, physical, chemical, biological, or whatever, belongs to a type in virtue of its similarity to other members of the same type, the ultimate extension of species-responsible behavior is responsibility to species other than organisms, the widest of all commitments. It may be formulated as follows.

An individual exists in virtue of those members of other species which do not, and to whom through his own existence he is responsible.

This is the principle of cosmic type responsibility. Whatever exists is a sample of all that has existed and all that will. Nothing is entirely alone in the universe and nothing is independent. The long-range self makes demands upon the individual that can be met only by his conceived identification with the whole of existence. No steps can be omitted: identification with himself as a whole leads to identification with his society, and through his society with humanity and finally with the universe.

Cosmic confrontation may be defined as *conduct toward any species in consideration of all other species.*

All encounters with all objects are included, the conduct of any material thing in consideration of all other material things. What exists is to that extent good, but the good obtained in this way (and all good must be) is to be had only at a price. There is a sort of excessive behavior to which the very existence of anything — the individual included — commits it, for in terms of the given date and place at which anything whatsoever exists, nothing else can. Some kind of usurpation is inevitable from the very nature of existence itself. At the same time it would appear that the highest function of the individual is to defend human and cosmic values in the midst of his social environment.

The moral, the beautiful and the logical, all are descriptions of the contents of reality, inventories of the profusion of formed materials scattered throughout the cosmos. The substance of the material universe is made up of the values of the good and the beautiful, hung upon the structure of the true. Matter has both truth and value because it occupies both space and time. If we knew the totality of truths about any particular material thing, we could deduce from this its value. Understood in this way, truth is a function of space and value a function of time, admittedly a contention requiring further elucidation.

The things to which a statement refers, and whose existence in the condition which is called for by the statement makes the statement true, could exist anywhere throughout the whole of space. To say, then, that truth is a function of space is to speak in an accurate but very general way. If 'true' means 'occupying space', then a true statement must be a general one to which no exception can be found. Thus the statement 'All men are mortal' means that 'All members of the species *Homo sapiens* who ever have occupied space, who do so now, together with those who ever will, must eventually die'. Truth is a function fo space because there are always vastly more absent than present material things in any given class.

Whatever exists in space must exist also in time. However brief the time, it must be of some duration: to exist is to endure. Thus endurance is a *condition* of value but it is not a *measure* of value. A rock last longer than a man but may be worth less. But generally speaking for anything to have value it must exist, and the longer it exists the longer it has value. Time on the whole is a rough evaluator, but it is not absolute and instead employs statistical trends. Value is a function of time even though it often happens that some things which are entitled to exist longer because of their greater value do not do so because of the effects of extraneous events and influences. It has been said that 'the good die young', but so do just as many of the bad, to say nothing of the beautiful and the ugly.

CHAPTER XI

MORALITY: THE BAD

1. Immorality

In chapter VII, above, I discussed human perversity as one result of an advancing technology, and pointed to some of the disorder it occasioned. That was an exceptional view, and so in the chapter before this one I endeavored to lay out the whole spectrum of human behavior with respect to morality. There, however, I was occupied chiefly with the good. Human life when looked at in all of its aspects is neither simple nor entirely admirable. The human individual lives in a society which exists in the midst of other societies, and he is confronted with a number of conflicting challenges which together often amount to more variables than he can deal with to his advantage. He must do bad things in his pursuit of the good, and so we have not adequately considered morality until we have looked also at the bad.

The most important fact about the practical life is that it is a mixture of good and bad, and is so of necessity because goods conflict. Hence any actions taken by the individual may have bad as well as good effects. But there is more to it than that. Many events have more than one cause and almost all have multiple effects. In morality, as in medicine, the desired effect is called the effect, the undesirable ones are called side effects. An action which is good in itself may have bad side effects. It is often impossible to do good directly to someone without doing bad to someone or somthing else. The individual thus is often caught up in a moral dilemma.

Man is born neither good nor bad; he is born with needs and also with drives which continue after those needs have been reduced. The needs as such are good but what is needed may be bad, and so it is in the acts performed to reduce the needs that immorality may enter, as indeed it may enter again in the left-over drives.

It is not enough to have considered morality abstractly, for it happens that the human individual does not live in an abstract world; he lives in a concrete world of material reality in which there are always conflicting forces. The use of force is inescapable in concrete life, and it may take many forms, from the mild kind involved in persuasion to the violence of brutal aggression. The situation is not a simple one, however, for force operates in a concrete world which is governed by the principles the abstractions represent.

It is time now to look at the limitations of those principles in practice in order to see whether they are responsible for further principles which are peculiar to practice. Ideals apply only to aims and ambitions targeted at conditions the individual would like to see brought about. Practice has to do with the extent to which his behavior falls short of those ideals.

The behavior of the individual is morally bad when he acts against the realization of the best that is in him: against the concrete ideals of his society, against the welfare of humanity, against the order of the cosmos. Immorality does not stop there, however. In addition to the morally bad actions of the individual, there are the morally bad actions of societies, and we shall have to take a brief look at these also.

I define bad behavior as *action which is taken in conformity with a lesser over a greater good.*

Although 'evil' has been the name for the intrinsic quality of extremely bad behavior, I will not use the term here because it has a theological rather than an ethical connotation, and I am concerned at this point only with ethics.

The bad is the result of incompleteness, the disorderly actions which follow when wholes are not brought together, just as the ugly results from the inconsistency which fails in the same way with parts. A good thing (or action) is one which has an effect more good than bad. Every activity has bad as well as good effects, for each is, so to speak, just over the border from the other. Perhaps the most baffling yet irrefrangable of all the facts in the entire range of concrete morality is that the best and the worst have a mutual affinity. They are drawn toward each other, in ways which Aristotle perceived and Hegel proclaimed. This fact severely limits the perfectability of the actual good.

Of course the problem is aggravated by the fact that goods themselves conflict. If only one of a number may be selected, the others must be rejected, and the effect of a rejection is morally bad. Thus 'moral' is in practice a captive term, indicating as it does the approval of one particular morality and the disapproval of all others.

2. Individual Bad Behavior

Individual behavior is morally bad when the individuals acts against the best that is in him, against the concrete ideal of his culture, against the welfare of humanity, or against the order of the cosmos. A few words now on each of these.

(a) *Bad Individual Behavior Toward The Self*

When an individual acts to reduce his needs he does so in terms of the good for himself. This follows from the definition of the good as what is needed, and the requisite behavior which seeks to obtain it is also good. However, the needs, as we saw, are organ-specific, and what is good for some particular organ may not be good for the whole organism. The urge of the individual to exceed himself which I have mentioned before is an extension of the particular features of organs as the ends served by the whole organism, often to its detriment. Thus the drinking of water is good, not only for the kidneys but also for the entire man. But what is good in moderation may be bad in defect or excess: men die quickly enough of thirst, while pathological water-drinking is a harmful deviation.

It cannot be emphasized too strongly again that in man a drive does not end when the need is reduced; instead the accumulation of goods is sought far beyond their possible use. There are no territorial limits to man's acquisitive ambitions. As it happens there is some justification for this in the need for ultimate survival, for the reduction of which, I pointed out, the individual would have to dominate the whole universe.

One peculiarity of human nature is the way in which excessive behavior is reinforced by artifacts, which set no limit to their own use. Artifacts can change the morality of behavior. The most obvious example is the recent development of weapons of total destruction, the targeted ICBMs, which are capable of destroying all life in the western hemisphere even though not everyone responsible for their manufacture and control has this effect in mind. Excessive behavior is in fact aided and abetted by the inherently general nature of artifacts. Once a tool exists, the tendency is to use it whether it is needed or not. Motor cars are often driven and guns fired only because they exist. Languages are often universal in their references even when nobody intended this feature.

Individual needs are reduced by means of artifacts, seldom without them. The same provision which enables man to get somewhat ahead of his primary needs, and so be left free for other pursuits spelled out by the secondary needs has resulted in the production of civilizations but has also been responsible for the bad effects of excessive behavior. The very nature of artifacts encourages the mechanism of need-reduction to continue after the needs themselves have been reduced, and so one of the side effects of artifacts is to reinforce excessive behavior.

The corruption of confrontation for the individual means in action, *'Conduct toward all others, in consideration of oneself alone'*. In bad behavior there is a lack of consideration of others in the conduct toward confronted objects. Bad behavior is always one-dimensional and single-minded. There are always individual faults, nobody is perfect. One by one we are brought low, as George Meredith observed in his poem-sequence *Modern Love*, 'betrayed by what is false within'.

Bad individual behavior is bad behavior directed toward the self. Foremost perhaps is the direct aggression against the self, such as self-abuse, self-laceration, self-betrayal, suicide. This statement requires qualification, for in some societies, the ancient Japanese for instance, suicide was considered an honorable course, a judgment made, however, from the social point of view; from the narrow point of view of the self, suicide would still have to be an absolute bad.

Subjectivity is the outlook characteristic of the morally bad individual. It is found at its most intense in certain types of mental illness. It runs there all the way from a concern with the self in simple unfocused anxiety to the solipsism of the catatonic schizophrenic. It is found to some extent in most average persons, since the pathological in most cases is only an exaggeration of the norm. Bad behavior for the average individual leads to an exclusive preoccupation with self-realization, to conflicts in need-reduction, and to a concentration on the needs of the individual regardless of social, species or cosmic consequences. Bad individual behavior leads the individual to transgress the laws of his society and to commit misanthropic acts.

In a sense the criminal is understandable enough. Bad social behavior is built into the organic individual: he has to kill to eat, and it is only a short moral distance from killing other animals with whom he has many affinities to killing men who stand in his way by wanting the same objects of need-reduction.

(b) Bad Individual Behavior Toward Society

Socially bad individual behavior is bad behavior toward society. It is the individual's transgression of the laws of his society which constitutes the most familiar form of the morally bad. The morality of a society is embodied in its established code of behavior. The morally bad, then, is that action by the individual which tends to reduce or cancel the effectiveness of the moral code. Bad social behavior is what is dealt with in the courts. It occurs when an individual seeks to reduce his needs in ways proscribed by his society, and is the commonest form of socially bad individual behavior.

Crime, it might also be said, is no respecter of social systems. No matter what kind of government is established, a monarchy, a democracy, a communist dictatorship, there will always be found individuals intent on subverting it. Strictly speaking, and in this context only, the worth of the government is not at issue. What is at issue is its endeavor to maintain itself against the bad individual's effort to undermine it or to benefit himself at its expense. Criminals are so common that it is fair to wonder whether any social order can ever be total.

The larger the population and the greater its concentration in cities, the more prevalent is the criminal element. The central problem in every modern civilization is the preservation of civil order, to assist which a complex criminal justice system had to be devised and maintained, no easy task given the multiplicity of other problems.

There have been many studies of the criminal mind, none of them conclusive. Certainly it appears that in an ideal society, if it ever existed, there would be individuals intent on its undoing. Whether this fact represents a failure of the ability to measure up to what is expected of most people or expresses a feeling of superiority to them, the result is the same: an individual who has to be branded a criminal because of his actions.

(c) Bad Individual Behavior Toward Humanity

There is a morality for all of humanity, and it is possible for an individual to act in conformity with it while doing something the majority in his own society recognizes as wrong according to its prevailing code. A German who tried to save German Jews was acting in the name of humanity but against the prevailing Nazi morality. What must have been the feelings of the Turkish soldiers who were ordered to dynamite the Parthenon or of those Christians who helped to destroy the Library at Alexandria? Did they not share in these acts of immorality toward humanity perhaps even without realizing it?

Directed destructive aggression is, alas, only too common. Individuals released from the moral obligations of their own societies by conditions imposed during wartime show nothing but ferocity toward the captured enemy, soldier and civilian alike. The massacre of prisoners taken in battle occurs more often than not. It was Genghis Khan's ordinary procedure to slaughter his enemies when his forces were conquering most of northern Asia and eastern Europe. The early Christians destroyed much of Roman culture before they learned how to use it, and the later Christian emperors of Rome and Byzantium did not show the forebearance their creed required. If there is such a thing as a common humanity, it is often submerged in a welter of more violent feelings.

(d) Bad Individual Behavior Toward The Cosmos

Man's urge to exceed himself accounts for much of what is most wonderful and terrible in human life. His monumental efforts to become one with the source of his feelings of sublimity, on the positive side, and to satisfy his frightful appetite for total destruction, on the other, both stem from it. He makes all-out responses to cosmic tropisms which often are superb in their effects but just as often are ghastly and horrible. The sun led Plotinus to mystical flights of religious ecstasy in solitude. That same sun led the Aztecs to the ritual performance of the bloodiest of human sacrifices.

In corruption of the fine arts and the pure sciences there is morally bad behavior toward the cosmos. The corruption of the fine arts for which individual artists are responsible consists in the pop arts with their lowering of the public taste, which in turn debases the quality of society. The corruption of the pure sciences is to be found in the promulgation of such false sciences as alchemy and astrology which promote false knowledge and lead to vain hopes.

I call behavior toward the cosmos bad whenever any destruction for which individual man is responsible occurs in the universe. Bad cosmic behavior is bound to result from the inability of anyone to act without disturbing his environment. If it is true that every material thing has a right to its own existence regardless of how this affects the existence of other things, then to protect its existence it has to fend them off at whatever cost when by their very proximity they impinge upon its integrity. Thus intolerance of opposition is a characteristic of material things.

In addition to the bad behavior of the individual, which has been my first concern, there is also the bad behavior of society. It can be directed (a) toward individuals, (b) toward other societies, and (c) toward other species. We shall have to look briefly at these, too, if we are to understand morality.

3. Bad Social Behavior

(a) *Bad Social Behavior Toward Individuals*

A good society would be one in which in addition to a concern for the welfare of its individuals there was a devotion first to truth and secondly to the acquisition and use of knowledge. Remove either and the result is bad. The Soviet Union is an example of a society which has the second but not the first. Perhaps the closest in of socially bad behavior toward individuals is the behavior of the Nazis just before and during world war II when they constructed and operated extermination camps for those of their nationals they regarded as undesirable.

Whewell, in his nineteenth century essay on morality justified the destruction of 'savages' by a 'Civilized State' (1). But perhaps the extreme of socially bad behavior toward individuals is exemplified by cannibalism, which is practiced still by some primitive societies and by nearly all societies when at war.

It is difficult sometimes to remember that the institution of slavery was common throughout most of human history. The most powerful nations practiced it, the most intelligent individuals condoned it. It was the conventional procedure in ancient Greece and Rome, it was taken for granted by Plato and had the approval of Aristotle. Only in recent centuries has it been condemned.

Respect for the life of the individual is a sentiment which is still limited to a very few countries, chiefly those in western Europe and the Americas. It is yet to spread to Asia and Africa. A government, like that of the Soviet Union, which has, and worse still, exercises, the option of imprisonment or death, for those who dare to criticize or differ with it in any way, certainly does exhibit bad social behavior toward the individual in an extreme form.

1 William Whewell, *The Elements of Morality, Including Polity* (Cambridge 1864, Deighton Bell), p. 401.

(b) *Bad Social Behavior Toward Other Societies*

War may be the commonest of all kinds of bad behavior. Most of human history is devoted to the melancholy record of conflicts between tribes, between nations, and between groups of nations. Wars are inevitable so long as the interests of societies conflict. Since societies consist in (among other things) human individuals, those individuals get caught up in the struggle between societies and so once again become involved in bad behavior.

World wars are the most destructive of all examples of bad social behavior. Millions were slaughtered on both sides in the last two and we may not have seen the last. Because of the destructive power of nuclear weapons, the next one could bring an end to the entire human species. Perhaps men will not be sufficiently mad to fight world war III, but their recent behavior in a continuing series of lesser wars does not favor that view.

Massacres are as old as history: both Sumerians and Egyptians were guilty of the razing of cities. Saul and David slew their thousands, and the Thirty Years War in Europe was as cruel and destructive as any, sparing no women or children. Coming down to our own times, the Bulgarian massacres of Armenians in 1894-5 and the Turkish massacres of those same peoples in 1914-15, the Nazi holocaust of 1939-45, in Angola and Chad, in the Central African Republic, in Nigeria and Uganda, a most impressive list, grim evidence that the hopes of the French Enlightenment were to be short-lived.

(c) *Bad Social Behavior Toward Other Species*

Human behavior is socially bad when the human species has a deleterious effect upon other organic species. I have mentioned that inter-specific aggression must be continued as long as men find it necessary to kill for food. Nature is cruel, and the last cruelty is the imposition of death which must happen to all so long as man remains the greatest of all predators. Any species can suffer destruction at his hands, and destruction as such is bad.

Thus the latest and most severe of socially bad species behavior must be attributed to man. I know of no other organic species which has destroyed so many of the species upon which it feeds. The balance in nature insures that a limited destruction permits the continuance of the lesser species and thus of course also the destruction. But thanks to man some species have become extinct, probably because of that built-in excess in human behavior that does not allow it to stop the drives once they are reduced. Hunters often kill for the sheer pleasure of it far beyond the demands put upon them by hunger.

4. Moral Strategy

The human individual who in the midst of his life proposes to consider the foundations of his conduct will understand immediately that he is obliged to distinguish between three separate domains: the ideal, the actual and the strategic. Once having by means of an ideal set the goal for moral actions, he must recognize that he starts from an awkward position, for he finds himself irrevocably involved in the field of the mixed. His chosen ideal has associated with it the compulsion of a duty, but he stands in the concrete world of good and bad as it exists both within himself and within his society. Being human he must operate in terms of a logic based on reasons learned beforehand and established in his unconscious from where it automatically dictates his choices. Moral strategy is what he needs to put the ideal into effect.

I have pointed out that there are moral commitments which are individual, social, human and cosmic. The obligations corresponding to these are: for the individual, to achieve self-transcendence or super-identification; for society, to embrace all of mankind; for humanity, to develop a higher species; and for the cosmos to order all species. Considerations of space have compelled me in the short brief compass of these two chapters to deal chiefly with the moral commitments of the individual. I did not attempt to deal with most of the others, but I have said enough to disclose the complexity which prevails at even the lowest of the ethical integrative levels.

Above the individual, then, hovers the choice of ethical ideals; before him stands the rich world of fact; upon him is the necessity for acting against the background of the latter and in the direction of the former. He acts, in short, by manipulating the facts in terms of some conception of the ideal.

Ours is a two-tiered world of the ideal over the actual. There is almost no one who is at home in both worlds. When the individual is an idealist he has not the slightest conception of how bad things are in the actual world or to what extent force is prevalent; and when he is a materialist he is not aware of the extent to which logic always manages in the end to assert itself in terms of the organization of the good and the beautiful. Both types are defeated by the fact that strategy involves them in utilizing the relations between the two tiers.

Of if only in all of a man's life the 'facts' were truly facts and his chosen ideal truly The Ideal! Instead he finds his situation in all three domains, the ideal, the actual and the strategic, to be insecure, tempestuous and approximate. They may represent the best that anyone can expect at the time and under the given circumstances, yet they do not prove that no absolutes exist. There is always the one that all of us are looking for among the debris of fact and among the choice of ideals.

Moral action involves making a selection of the moral integrative levels; yet moral action is not itself a stage-process, for all activity is usually singular and one-dimensional. The principle of selection in moral action may be stated somewhat as follows.

Try to act so that the effects of your actions satisfy the ethical demands on a given moral level in such a way as to enable you to reach upward toward higher levels.

The higher an individual is able to rise in this way through a majority of his confrontations, the more of an individual he becomes, that is to say, the closer he approximates to entire man. Obviously no reconciliation of conflicts can be reached through any compromise. Here it is necessary to invoke the approved use of excessive behavior in moral strategy, for only the best is good enough, and the best is not a mean between extremes. For the extremes themselves, the method of over-reaching is required. What is in the background always, conditioning all moves at the practical level, is the definition of excellence as the quality of perfection. Such a quality can be experienced in little and need not be put off to ultimates.

I have been describing the active phase of moral behavior. There is a passive phase also, and it may be characterized in general as moral detachment. Its central concept is *standpointlessness*, which involves the method of employing general empirical knowledge to rise to an extended view of existence in order to understand and to feel at one with as much of it as possible. It is natural rather than super-natural, and it is aided through intuition by the arts, through feeling by the religions, through controlled activity by the sciences and through understanding by the philosophies; all aimed at the same goal.

Standpointlessness is the ultimate strategy of cosmic obligation, and it too is a process involving four stages. In the first stage the good is preferred over the bad; in the second there is acceptance of the bad as well as the good; in the third there is a love of difference; and in the fourth we reach a process of progressive detachment. Standpointlessness requires the definition of reality as equality of being, and the recognition of that principle in every material particular. Cosmic morality incorporates a number of features: equality of goodness; cosmic solidarity; pan-provincial caring; and the metagalaxy as the ultimate object. At the end of all moral striving the actor ought to be replaced by the spectator, with a feeling of sublimity as the final reward of standpointlessness.

All individuals are ceaselessly involved in action, and so a dual attachment emerges. It is not advisable to live in two worlds, the actual and the ideal; but short of a fatal cleavage and an exclusive preference for the one over the other, there is no better alternative. This ambivalence makes of the moral life a delicate affair similar to the task of the tightrope walker, and transforms every moment into a quivering drama, with the triad of balance, compromise and unification a necessity as well as a goal. This is the demand put upon the individual who is profoundly human, that he make every decision driven by the organization of the entire man, a decision always based upon moral grounds that are deeply imbedded in him and imposed upon him by the whole nature of his being.

CHAPTER XII

RHETORIC

1. Languages As Artifacts

Before introducing the topic of rhetoric, I need to remind my readers of a few points which were established earlier about the behavior of man as an organism.

All organisms aim at survival, and man is no exception; but in addition he has extended his aim quantitatively until it amounts to a qualitatively different affair, for he aims also at ultimate survival. He is called on then, to deal adequately with the environment, since he must dominate it sufficiently to be able to make it over into something more serviceable to both aims.

Understood in these terms, human behavior may be described as the result of a continuing series of interactions between man and materials, with the man undertaking to alter the materials but just as often being altered by them. Usually, language is involved in this round of interactions, for linguistic expressions is the form in which beliefs are stored in the memory as engrams, and actions taken largely on the basis of beliefs. Thus we may see here how language functions as a material tool.

Languages are of course social inventions. The individual devises them in cooperation with his fellows, and the patterns have to be stored and communicated. A communication does not become a language until it achieves the symmetry of a two-way operation. Unless the sender of a message could exchange roles with the receiver on the occasion of the next message, it is not legitimate to call a mechanism of communication a language. Thus the chemical code of information in the genetic material of organisms which instructs the next generation how to make a replica of its parents is not a language although it is a communication. Both symbolic communication systems and languages can be seen to be necessities of survival. The most familiar of such systems are the colloquial languages, the 'natural' languages which have developed in the course of the life of communities.

Let me begin with a more primitive account. Survival is possible for the individual only if he comes to terms with his immediate environment, meeting its demands on him and extracting from it what he needs. Both activities involve interaction. The stimuli to language-using are the material things in his immediate environment which offer need-reductions or threaten injury. Continual encounters with these or like objects (members of the same class which can be readily identified) call for naming in order to facilitate recognition upon subsequent occasions.

Language arose as a reponse to these interactions. Recognizing objects involves naming, and referring to them to make it possible to deal socially with tasks too large for the individual alone, requires the making of sentences by combining names. Thus language, which consists in signs made by shaped sound waves or by scratches on hard surfaces, is a set of material tools invented in order to deal with other materials.

Earlier I made the point that the 'material' is not the 'physical' because the physical is the first level of organization of materials. All organizations of matter above the physical, such as the chemical, the biological, the psychological and the cultural, have their own names. Since these integrative levels of organization are cumulative upward, there are no non-physical materials, but we call those materials which have merely the physical properties 'physical'. Thus matter is a neutral term but is always at least physical.

Language, we can now say, is a material tool specifically employed at the psychological and cultural levels where it functions as a conventionally established signalling system. Language in use consists in chains of discretely coded signals which pass through material channels of transmission and are then decoded. The channels may be naturally occurring materials, such as air, or artifically contrived ones, such as books or telephone lines.

Wittgenstein in his *Tractatus* has described how he thought language was constructed. Material things, he said, are named and the names combined in elementary propositions. The elementary propositions are then combined in complex propositions, and the multiplication of complex propositions makes up a fine mesh network which finally mirrors the world. Thus language is called into existence to describe and to deal with material things.

No doubt such constructivism does occur, but there are other approaches to the same method, for instance the scientific method of inductions from observations to hypotheses which are then measured against relevant portions of the world, and of deductions made from axioms and then applied to the world as a test of their representation. But all fo this involves relations with materials which still constitute the criteria of truth or falsity. An ordinary language is an open system in almost every respect. Its syntax is continually undergoing revision, and its vocabulary is constantly being altered as words change their meanings and flow into or out of it.

Language fascinates philosophers because of its complex and elusive nature, but there is nothing mysterious about that part of material culture in which communication blends with storage for purposes of control; it is a tool like any other, even though a pervasive and indispensable one and so an important kind of technology.

As I have noted in this book now a number of times, an artifact is a material altered through human agency, and there are two kinds of artifacts: materials shapted into tools, and languages. In this sense a building is just as much a tool as a bulldozer, a violin as much as a paved street. This is quite conventional usage, but it may seem a bit more unfamiliar to say that a language is also a tool. However, language too uses shaped materials, at one time baked clay, then papyrus reeds, palm fonds, strips of bamboo, wood, stone and leather and now of course paper. The symbols at first were pictures and only later were replaced by abstract signs to which conventional meanings were attached.

It is by means of those material objects called signs that men are able to refer to other material objects. All language points to material things, directly or indirectly. If directly, then by specifying what is true of the class; if indirectly, then by referring to other linguistic expressions which eventually refer to material things directly.

A language, is a set of signs together with the rules for combining them in order to make them into indicators of material things or events. Often expressions in a language refer to other expressions, but eventually there is a material reference for the second expression or for the third, or for one somewhere down the line. It can be shown that eventually all signs point to materials or their properties.

The capacity of some expressions in a language for referring to other expressions was what opened up the references to a second tier of abstractions over the first tier of facts (1). Through the centuries what was written down was often thought worthy of preserving for use by another generation. This involved additional procedures providing store houses or libraries, and made necessary a higher degree of social cooperation. Thus there arose an entire technology of knowledge with its need for more complex channels of communication. Perhaps we should look next at the history of this development.

2. The Origins of Formal Communication

It has been the dominant thesis of this work that throughout the life of mankind, in fact from its very beginnings in prehistory, man has assisted himself in his practical tasks by resorting to the aid of technology. This is no less true of communication.

1   The abstractions have already been discussed under the title of universals in chapter V.

until Aristotle and Isocrates formalized it and means of principles and the latter by exam- cient Greece meant public speaking. Aristotle rving in any given case the available means of rules for making public speeches effective. d most citizens were present at all important ave been easy to influence Athenian society by earshot of the speaker's platform. For the rest ual form.

rary, Isocrates who took the process one step rhetoric into a formal discipline, and by teach- rival the influence of Plato himself. Because of vered his speeches, instead he wrote them out, f what forms persuasion can take. The effect of r beyond that, but we are not here concerned his influence as a rhetorician. There is no doubt ed influenced all Roman public speakers. It was at the Romans inherited, and it went beyond ation, which is the art of communicating to the

f the essentials of grammar and syntax, consisted ans of examples and imitations. The practice is a precedent in the last quarter of the third dian conquerors of Sumeria studied and imitated ey regarded as classics, but the knowledge of ks. For the most part the Greek inheritance as not include science and mathematics. and in the period of the Christian domination of both learning and education were neglected. in monasteries, but it was not available to the uch thing as general education. Its appearance greatest increase was in the nineteenth century special forms in recent decades.

hnology

ducation for all the members of society, was n of printing. Johannes Gutenberg in 1456 with his edition of the vulgate Bible printed on movable type showed that books could be manufactured at a cost which put them within the reach of many who would not have been able to buy the relatively expensive copies which before that had to be made slowly by hand.

The new technology introduced immense changes, including universal knowledge and a growing sense of individual importance. Universal suffrage was bound to follow. After 1822 the printing press, mechanical typesetting, mechanical binding, the hand lever press, the cylinder press, and the flat bed machine, followed one another in quick succession.

Indeed it could be argued that Gutenberg's invention eventually was responsible for mass literacy and with it popular democracy. With books so accessible, there was an urgency for everyone to learn how te read, and with reading came thinking and that inevitable sense of responsibility which led to the demand for universal suffrage.

Gutenberg's invention of printing also marked the doom of oratory for awhile. With pamphlets and books it was possible to reach many more people. Men sought to influence each other with broadsides and periodicals. Public speaking became a desultory affair.

This situation was abruptly altered in our own day by two inventions: first the radio and then television. By means of the radio the voice of a speaker could be amplified and distributed far beyond anything that had been contemplated before. Greek politicians were able to appeal personally to all of the electorate in open meetings, but that advantage was lost when populations began to increase.

Probably President Franklin Roosevelt's many terms in office were due to the persuasiveness and reassurance of his voice on the radio, from which no listener among voters was immune. Proportionately speaking, we have gained more with network television, for politicians can now be seen and heard by the citizens of an entire nation.

The art — and the technique — of oratory were quickly revived. There is still no formal discipline; public speeches, even when they are carefully planned and written out to be read before microphones and cameras, are not given the strict form that Isocrates and Cicero had designed for them. The forceful effectiveness of the personality still counts for a great deal. But the point is that public speaking is once more a power in the world and television was responsible for the change.

It would not be possible to estimate the influence of technology on communication by reciting the history of the four great developments: writing, printing, radio and television, to say nothing of the description of other important additions. Communication has been intensified as well as extended. There has been an enormous growth in information processing systems of all kinds: large-scale electronic computing machines, networks for linking computers, semiconductors, satellite and groundbased microwave data communications, and now the prospect of further advances made possibly by the next step in miniaturization: VLSI, or very large-scale integration.

oserved that the earliest civilizations were accompanied
e defined as the communicating of signs by means of
is no doubt much earlier but has left no trace. Evi-
en found in the remains of the civilizations of ancient
ted to some five thousand years ago.
ting has survived. What remains tells a story of the
noble warriors to those featuring scribes. The aristo-
to fighting, relied upon a class of civil servants who
l needs of accounting and administration (1). Their
serve and disseminate religious dogmas or philosophical
indentations were mostly records of economic trans-
bes worked for temples or palaces. Writing was a
igh confined to a small professional class, made the
possible.
trade, and so there came about a formal process of
textbooks. Teachers' models and children's exercises
dating from the same period (2). The education of the
i its principles were not formulated until much later.
rhetoric; the attempt to influence people by means of
in much before the invention of political democracy by
ress of civilization was made possible by technological
t it did not move steadily; instead it was marked by

e history of Greece. The preliterate society still left its
s of Greek civilization. The books that have survived
ats of warrior heroes, as in Homer's two epics. The
f writing to the recording of abstract thought in Plato's
years, a rapid advance indeed!
overy of universals, though if we had more of the work
the credit might have gone to them. At all events it
perhaps the greatest ever, for abstractions have made
control of the material environment, though the Greeks
ught only of the control of human behavior through the

ation in Antiquity, G. Lamb trans. (London 1956, Sheed and

Begins at Sumer (New York 1959, Doubleday Anchor Books),

The conquest of space as well as of time by communication systems brings with it special problems. Technological improvements in communication would have to include the immense influence of the telephone, which has extended the range of the speaking voice until it can now circle the earth. Global communications recently received the aid of satellites to assist telecommunications. Newspapers are simultaneously printed in a number of American cities, and were it not for national and linguistic boundaries this could be done world-wide. For the fact is that communication and - I might add trans- portation - have developed more quickly and efficiently than political associations. If these two technological advantages are to be given their true worth, then nothing less is called for than a global state.

In short, technology has transformed the method of communication and with it the entire life of human society and culture. And this is true not only of social relations in general but also the relations of the individual in particular. Communication prepares him for participation in his culture: he acquires a language when he is young, and learns the most effective ways to use it when he grows older.

Communication is by no means all verbal. Recently there have been the beginnings of the study of body language (1). Meanings are often transferred from one individual to another through physical movements such as facial expressions, hand and arm gestures, or even inclinations of the torso. The study has not yet been reduced to a set of prin- ciples and exercises, but no doubt in time it will be.

Every forceful personality involves the use of body language though in a much less obvious way. When people in the theater refer to an actor's stage presence, this is what they are talking about, the power some individuals have of moving so that those around them feel the power of their personality. Actors, politicians, and to a lesser extent others, are able also to communicate messages by bodily movements.

4. Information Theory

Language may be considered a broad name for communication theory. Under this head- ing it joins communication engineering and information theory, with the emphasis on its technology and on the material media rather than on the message.

The information theoretic analysis of language has disclosed some interesting facts about it. The individual is a source, his formulation in language of the message he wishes to send is his version of an encoding transmitter, and the person who hears him or who receives his writing is a decoding receiver. There are no messages without a medium: materials of some sort must be used, usually a channel which has been devised

1 Julius Fast, *Body Language* (Philadelphia & New York, 1970, M. Evans & Co., distributed by J.B. Lippincott Co.). See also Desmond Morris et al., *Gestures: Their Origin and Distribution* (London 1979, Cape).

for the purpose. The construction of a message is part of communication engineering, and the effects of its contents a kind of human engineering — what Edward L. Bernays once called *The Engineering of Consent* (1).

Some of the greatest advances in knowledge have come from the study not of natural objects but of artifacts. For instance, the science of thermodynamics was suggested by the steam engine, aerodynamics by the behavior of the airplane, and the theory of communication by the telegraph. Similarly, information theory issued from the study of electronics and magnetism.

Rhetoric is the older term, the newer one has to do with coding, channels, amount of information, entropy. Rhetoric has to do with ordinary language, with the speaker as the source, his framing of words as the coding, speech sounds or writing as the channel, hearing or reading as the decoder, and finally comprehending as the recipient. In the broadest sense, any transfer of energy may be described as a communication since it is a 'message' conveyed by signals from one body to another. In recent decades the 'bodies' involved include machines. The sending of messages from one machine in which they are encoded to another in which they are decoded has become an elaborate and complex affair.

Information theory is perfectly general. It considers such questions as how many signals can be sent over a channel and how fast; how much the channel itself interferes, that is to say, how much 'noise' it contributes. It deals with such absolute considerations as channel capacity and the number of digits required to transmit information from a particular source. Engineers know for instance how many bits per second of errorless transmission it is possible to send over a given channel despite the level of interference or 'noise'.

However, colloquial English is quite another matter, and offers so many possible interpretations using precisely the same arrangement of words that encoding becomes a serious even if not an insuperable problem. Often eventually the meaning of a sentence can be interpreted only by tracing it to may elements in the world inhabited by those using it.

The value of a method of communication is the extent to which it serves as a useful mechanism of transmission without too much distortion of the contents transmitted. Information theory is clearly an advance made possible by the new techniques devised for the use of electrical communication systems. If it is efficient it does not appreciably alter the messages but rather extends their range. This may be the first time that anyone has considered the transmission of messages quite apart from their contents, and it has been very influential.

1   (University of Oklahoma Press, Norman, Oklahoma 1955).

5. The Colloquial Languages

The symbolic nature of language is hardly a topic that anyone could rightly describe as neglected, but the same cannot be said for its material nature. It has been considered negligible where it has been acknowledged at all. McLuhan's point that 'the medium influences the message', however, is just beginning to be understood. It is well known for instance that telegrams constrict language, and that the printing press elaborates it. A skilful communicator, however, makes words say what he wants them to say, and not what they want to say or what their medium of expression inclines them to say.

No doubt the medium does influence the message, but there is another, and in many ways more serious, type of distortion, and it comes not from the medium but from the anticipated character of the decoding receiver. The rule for this is a simple one. As a result of observations made to date it is fair to say that the more far-reaching the media, the more trivial the message. Radio and television have developed no Homers, no Shakespeares, no Bachs.

The mass media are apt to drag down the information which comes over them to a level which is intelligible to those whose intellectual equipment is not very great. The same is true of most newspapers and popular magazines, and even books: the greater the immediate circulation the more superficial the contents. This is not an absolute rule, of course, but it is indicative of a strong statistical trend.

It can be argued that such a development is the fault of those in charge of the media, or of the system whereby the media are financially supported. So long as commercial advertising firms or governments which rest on the support of the masses must furnish the funds, the appeal will be to the greatest number of people. More definitely, then, it is the fault of the system, which should somehow be supplemented.

The ordinary or colloquial languages themselves are the chief offenders. They are collections of artifacts which are combined uneasily and loosely to make up an instrument of communication, and they contain all sorts of oddities. For instance, no one has ever used an entire language, only segments of it. It is to be doubted whether in fact anyone ever mastered a whole language, and so the individual is always in the strange condition of relying heavily upon an instrument of which he knows some parts only, and out of this knowledge must select the small finite combinations he wishes to use. There are some half million words in the English language, but a sentence, which is the unit of expression, seldom contains more than a few dozen.

This is one peculiarity; there is another. Most tools have been made as the result of deliberate efforts, but in the colloquial languages we have the spectacle of tools which grew up unplanned over the generations, continually being enlarged and amended unofficially and unprofessionally. Some of the slang of the streets, which is often the work of illiterates, may pass permanently into common usage. For the fact is that anyone who speaks a language may influence it. The most unplanned and uncontrolled of human tools happens to be also the largest and most indispensable. There are no authorities in charge of colloquial languages, which operate freely and clearly.

The colloquial languages were not planned as wholes, but nevertheless they have properties as wholes. It was noted by A.N. Whitehead and others that philosophies are implicit in them, specifically in their grammar and vocabulary but more especially in their syntax. A philosophy in this sense is a system of total explanation, and it is so imbedded in the language that to learn the language is to accept the explanation. Such implicit dominant philosophies were not constructed deliberately; they evolved as the language itself evolves, until one day they made their presence felt as essential parts of everyday communication.

Speaking a language in any full sense of the phrase means taking for granted in the hearer or reader the same set of fundamental beliefs that exists in the speaker. Much more is encoded and sent out over the channels, received and decoded, than either the sender or the receiver is aware of knowing. Whitehead complained that to express a new philosophy in an old language which already has its own philosophy involves a conflict and a confusion that threatens intelligibility.

The point is that when we use an ordinary language we inadvertently include in its use some extraordinary factors. In principle, a proposition should not be accepted by belief without examination, and the more fundamental the belief the more careful the examination. For we do act from belief, and the gravity of the situation is intensified when we remember that we act crucially and critically from our most fundamental beliefs. And so we should be extremely careful of what we admit to the status of belief. In this direction ordinary language is beguiling, and slips over on us much of which we have not examined and would not have accepted.

The almost religious faith in the efficacy of ordinary language held by Austin, Ryle and their followers is a consequence of the Scottish school of common sense given a linguistic turn by G.E. Moore and Wittgenstein. It is logical to assume that ordinary language is the natural vehicle for the expression of common sense, which is the ancient and long established set of beliefs of the members of a particular culture. Until the astonishing discoveries of experimental science in the late nineteenth century, common sense served very well, but now it may be downright misleading. Science has given a new dimension to natural knowledge.

The trouble with ordinary language is that it no longer embodies the entire breadth of human experience. It is not adequate to describe what the scientist learns in his laboratory or what he undergoes in his space craft. For these and similar experiences he requires the technical languages he has invented for the purpose, including the complexities of mathematics.

One peculiarity of technology is that in engineering it is never known who will be served. A bridge builder cannot tell who will cross on his bridge, how many people or for how long. The Romans who built the roads in Britain did not envisage the modern traffic which still uses them. This is not a critical problem and it does not have results which are crucial even though they are far-ranging.

But with the use of language as a material tool the situation is far otherwise. For in communication engineering there is the unusual situation that the message may be held in the channel indefinitely. We can still decode Egyptian hieroglyphics and Assyrian cuneiform. It is relatively easy to start a series of events but absolutely impossible to predict with any accuracy what effects they will have, how widespread these will extend in space and time, or how forceful they will be. The writer may be speaking to future generations many thousands of years removed from him and, as in the case of Greek philosophy for instance, have an immense and unforeseen effect on cultures which at the time could not even have been imagined.

6. Technology and The Use of Language

Language is that technology in which we attempt to describe and to deal a special way more with the similarities than with the differences we discover in material things. That is why language is so incurably general. The universal nature of language makes it difficult to define a single and unique particular, even though all particulars are unique to some extent. Russell and others found that in most instances if not in all, to name a particular is to name its class, and it is just possible that there may be no exceptions.

It is interesting that the elaborate technology of the material theory of reference by means of which we can designate material things was the outcome of the practical exigencies of dealing with the immediate environment in order to reduce our many organic needs. Sometimes language conveys reports of how to deal with materials in a way which bring them into conformity with the reduction of those needs. Formulas, the equations of applied science and technology, blueprints, recipes, textbooks, these are clear instances. They are all in a certain sense engineering directions. Thus while the conveyance of language is always an affair of physical technology, it is different with meaning. But, as we have noted, the meaning, too, is material. Propositions may be true or false, but so long as they are known to be true or their degree of truth or falsity remains unknown, they are accepted by belief.

No doubt language influences action, but it is also true that action influences language. Because of the traditional emphasis on reason and sense perception, the role of the musculature and of all the activities which are involved in the derivation of reliable knowledge must be included in epistemology.

Let us come at the technological problem from another perspective. Communication is social, and the construction of society does not consist merely of human individuals but must include artifacts. If we remember that ordinary languages are artificial constructions (in the sense that they consist in altered materials), then it is true that any relation between two human individuals is almost always mediated by at least one artifact. Materials altered through human agency furnish the cement by means of which human individuals are brought together and made parts of social institutions. Thus the more developed the technology the more advanced the institution. Technology is the driving force of human culture.

Language is seldom used without other tools, and even more rarely are other tools used without language. The speculations carried on in science and technology deal with the similarities we suspect to exist in material things. Discoveries and inventions in the design and control of some of the potentialities of materials are always conducted in terms of a language of some sort.

Great delicacy is required to be sure that the instrument by means of which communication is undertaken remains a means and does not interfere overly much with the message. Theoretically, the meaning of a sentence is the same whether it is spoken or written if the same expressions in the same language are used, but with regard to the original meaning a translation from play to novel or from novel to play more often than not involves a change. We ought therefore to talk not merely about languages – they do not exist in isolation but always are carried by material media – but about the material media themselves, in a word about the relation between meaning and the mechanisms of communication.

In considerations of meaning, denotation must be sharply distinguished from connotation. Words and phrases become encrusted with meanings through repeated and prolonged usage, and so there arises the complication of connotations added to denotations. In direct communication, say a statistical report, denotation is uppermost and connotation minimal if it exists at all. In direct communication, say a short story, connotation is uppermost and denotation minimal.

But let us confine our attention to direct communication. Here, the uncertainty of the meaning of a sentence is increased in proportion to the number of times it is used. If the sentence is widely applicable it will be used frequently in connection with different kinds of occasions, and so it will acquire a number of connotations and they will increase the uncertainty of its denotation. For the greater the number of possible meanings the greater the degree of randomness, and so the number of possible meanings decreases the probability of successful decoding. Thus the entropy of the meaning of a sentence is a function of the frequency of its use.

In this way technology enters into the question of meaning. Since the meanings of material things change, the language changes with them and there are no fixed meanings; what was decoded might differ from what was encoded. Mace was once a weapon of war, then it became a symbol of sovereignty, next it served as a ceremonial staff of office borne before designated officials, and now it is an expression of academic status. The meaning of a message will be altered in ways not indicated by its source. Communication is more damaging when it is thought to exist and does not than when it is thought not to and does.

Wittgenstein, in his second book, the *Investigations*, turned against language and charged it with obscuring our view of the material world. He proposed that a close analysis of ordinary language would reveal the degree of distortion, a knowledge which we would need in order to dispel it and thus come to an unimpeded set of perceptions. But in all this there is an assumed correspondence theory of truth of a special sort: that when language correctly represents the material world, then and only then is it able to speak for itself.

It is only too often forgotten that in a certain sense this goal has been reached. Wittgenstein's comments apply only to ordinary or colloquial languages. They do not apply to those special languages used by the experimental sciences, where new words are assigned specific meanings and underscored by mathematics. The scientific languages are not as rich as the colloquial languages, they are for instance entirely lacking in connotaions at this stage of the proceedings; but they certainly are accurate and unambiguous, and so are subject to none of the subtleties of interpretation for which Wittgenstein and his followers are justly famous.

The problem is to build a communication system in which the meanings can be standardized by constructing channels which do not allow appreciable alterations in the message. This might require a direct attack on the rigidities of language. The more subtle the ideas to be communicated the more flexible the required language. Deliberate modifications of ordinary language have seldom if ever been undertaken, but they are not beyond the bounds of possibility. It is necessary only to remember the technological aspects of language in order to remind ourselves that language itself, together with everything that it conveys, is a property of matter.

CHAPTER XIII

POLITICS

1. The Factor of Technology

It is fitting that a chapter on politics should follow one on communication. For the violent political changes and prospects that we are all witnessing today are the result of progress in communication and, I should add, also of transportation. It would be diffi- cult to see how the old social and cultural divisions will continue to suffice when we note the integrating changes that are taking place. A world-wide trade has rapidly accelerated, and so has travel. Messages can now be transmitted around the globe in a few seconds and men themselves can be carried the same distance in less than a day. Thanks to these two developments, the world is a smaller place, and so calls for a corresponding organization.

There is no corner of individual and social life which has not been influenced by technology. From the control of human population to the use of artificial intelligence, from the expansive knowledge about the universe to the intimate knowledge of processes in the cell, technologists continue to make advances affecting the kind of people we will be and the kind of environment we will have.

Small wonder, then, that technology interacts with politics. A new kind of material production inevitably brings with it a new form of rule. The rise of the factory system in England and Germany was responsible for capitalism and private enterprise, the reaction to it brought about communism. Even now the decisions of government influence scientific research and development, and determine the direction of invention and pro- duction, in a total way, as in the Soviet Union, or partially, as in the United States.

But even as communism spread, it became more and more evident that it was outdated. Its success in underdeveloped countries belied Marx's prediction that it would prevail first in the most advanced industrial states. If, as Marx himself said, politics is a product of the means of production, then it is true that now another political system is called for. The automated factory, computer operated, together with an entire spectrum of new inventions do not lend themselves to communism but may in time contribute to the need for a new kind of rule. Changes are occurring so rapidly that no one is able to predict just what form it could take.

We can now see that democracy provided the order appropriate to a free economy, and has worked admirably in that environment. It made possible a number of other developments that could not have been foreseen, such as the successes of the experimental sciences. For engineering it has provided the atmosphere so necessary to the flourishing of ingenuity and inventiveness. In an intensely important sphere it has produced and guarded the civil liberties, which have been responsible not only for the happiness of individuals but also for enlightened social progress.

But in another way democracy has dug its own grave: the very liberties which allowed for improvements in technology also permitted those political organizations which a weak central government was unable to control. It is only too often forgotten that the freedoms guaranteed by democracy allowed for the rise of communism. Marx and Engels lived and worked in England when other countries in Europe failed to provide the necessary refuge. By its very liberalism, democracy made possible the most serious challenge to it that has arisen since it was first established. In that sense, communism is a side-effect of democracy.

The result of these developments is that we are now in a period when two outdated systems of political organization are struggling for victory. Democracy cannot govern when economic and social problems call for a stronger central government. Communism does not offer the opportunities for progress that the technologists require; its control is too rigid and in most cases its leaders too ignorant to allow for the quality of leadership that individual imagination and initiation need if there is to be progress in every institution.

In short, if democracy is outmoded, so is communism, and in a very much shorter time. The problem this situation presents is a very clear one: how to have the kind of central political organization which will encourage individual initiative and civil liberties while establishing and maintaining social order. So far the solution is not clear.

2. Technology and Bureaucracy

Our knowledge of the integrative levels in nature is inverse to their complexity. The physical as the ground level must be the simplest, yet in our catalogue of information it ranks very high. We know more about its intricacies than we do about the others. This is a matter of ignorance, however, not a property of structure. If each integrative level above the physical is more complicated than the ones below, the chemical more than the physical and the biological more than the chemical, then obviously the social or cultural level, being the highest, must also be the most complex.

That clearly does not match our knowledge. Sociology and even many culture studies endeavor to analyze the elements of their level entirely in terms of human individuals and social organizations, as though these existed and managed their inter-relations without the use of artifacts and were simple affairs at best. But this would be entirely misleading. For it is not society which must be considered: the smallest social unit is a culture, and this consists in individuals, social organizations and their *artifacts*. An implicit dominant ontology runs through a number of institutions and constitutes their common bond, but this is far from being the whole story, for almost all social relations involve the use of material tools, including in that term the colloquial languages; but the analysis of the complexities of cultures has never been undertaken.

Cultures, in short, are organizations of men and the artifacts technology has been able to produce. Politics, as a subdivision of culture and one of its most important and indispensable institutions relies heavily upon technology. This has always been true, it is even more true now that governments are confronted with immense and growing populations. An elaborate technology is necessary to deal with the tasks of governing a population of 200 million which present problems unknown to the ancient Greeks.

With the rise of large populations and the rapid advance of technologies, artifacts of the kind intended to be employed only by large social groups are produced in greater number, and some only by nation-states. Telephone systems are good examples and so are capitol buildings, reservoir and dams, railways, airports and harbors, bridges and cyclotrons.

With the big artifacts comes an increase in the need for controls. The larger the country and the more industrialized, the greater the bureaucracy that will be needed to govern it. And bureaucracies operate almost entirely by means of the products of technology. Think of the number of computers, the filing and retrieval systems, available to the federal agencies in the United States, to say nothing of the additional ones employed at the state and local levels. Moreover, the larger the bureaucracy, the less efficient; bureaucracies are in this sense self-defeating, each one creates more problems than it solves as it outgrows its original function.

Bureaucracy, it might almost be claimed, represents an effort to serve the excesses of technology. As the proliferation of tools and instruments grows and their use increases, so do the rules and regulations governments feel constrained to put into effect; and as a consequence the bureaucracy expands to administer them.

Artifacts tend to proliferate when new uses are found for them. This has certainly happened with the interferometer and the radar antenna which made possible discoveries in areas not originally included in the range of these instruments. The interferometer was a device for testing the speed of light in relation to the motion of the earth, while radar was a practical machine for sending war messages; now both are employed in making astronomical discoveries in distant space. But just as often the employment of artifacts can be a detriment. It is not uncommon now to find a retail business employing computer-operated terminals when the volume of sales does not justify such an elaborate technology. The most extreme example is the war machine based on a nuclear arsenal.

It is a grave and open question whether all the people in the world, now numbered in the billions and increasing exponentially, will ever be governed together. Naturalists are fond of observing how animal populations are kept within bounds by predators, in an ecological balance that is able to maintain itself through the centuries and even the millennia. Yet it may not be a popular notion that war serves an ecological purpose. Every instinct of the humanists goes against the kind of procedure followed in the first war when the Germans on one side and the French and English on the other stood toe-to-toe on the Marne, and for four years slaughtered each other by the millions, yet if there were too many people, this may have made life for the survivors possible.

It is a striking but seldom acknowledged fact, that we do not have even the blue prints for an ideal society. We have no current guesses about what kind of government there should be. But we ought to acknowledge at least this: that the ideal state would include, in addition to the proper hierarchy of social organizations, plans for the production and employment of artifacts in proper proportions. The formulation of such an ideal remains an urgent necessity.

3. The Politics of Technology

Inequalities abound in nature. Lions eat whatever other animals they can catch, but there is no species which preys on lions. Among human individuals, inequalities of muscular and intellectual strengths are common. The attempt to extend obligations as well as privileges equally to all would produce a distortion; for everyone is not up to sharing obligations.

Equality, in other words, is an inadequate ideal, and all attempts to achieve it are doomed to failure. We have noted already that although communism overthrew the capitalists in many countries, it succeeded only in producing another privileged class of high Party officials.

A technology of luxury does exist and is highly prized, but it has not yet been extended to all. Everyone cannot own a large house, a Rolls Royce, enjoy a gourmet cuisine, and wear expensive clothes, though these are all material productions. Their very existence brings about a social inequality which the political system must find some way of accommodating. As long as such luxuries exist in a limited supply, there will always be a privileged class to take advantage of them.

A wholly egalitarian society, then, is not even an acceptable ideal. In the long run it must always prove impossible to legislate social inequalities out of existence; all the state can hope to do is hold them to a minimum. Absolute monarchies, in which economic differences have been carried to extremes of wealth and poverty, as for instance in France with Louis XIV, tend to breed revolutions. In the long run, however, even these succeed only in exchanging one privileged class for another.

The solution may lie in the control of technology, which can be designed to produce for the most part only what can be made available to all individuals, but there is no indication that the human species is ready for that. Perhaps another criterion is called for: a system in which those who make the largest contribution to society would receive the greatest rewards from it. The Communist Manifesto has to be rewritten: *To each in proportion to his unequal contribution, for each minimally because of the universal equality of needs.* Under this dispensation, no one would go hungry or lack shelter and medical attention, but great scientists and artists, indeed all of those whose work has been of the most benefit to society, would receive the largest material rewards. Everyone would have a basic minimum, while criminals might receive less because of their disvalue to society.

In every instance it is the management of technology that makes the difference. Marx talked about the means of production, but he never took as seriously the design of artifacts, even although what is produced must first be invented. The design state is the crucial, one for that is when social needs are anticipated; and it is precisely there that management must make its most important decisions. Indeed management becomes more intensely a political imperative as artifacts come to play an important and even crucial role in human life. Governments are installed and they are overthrown, but always in the interest of the production and distribution of artifacts.

4. Technology: The Negative Effects

The alternations of peace and war throughout history seems to guarantee an ambivalence, the same one, in fact, that we noted earlier in motivation. The urge to destroy as well as to build is repeated in every phase of cultural life. The increase in the complexities of social organization has been accompanied by a corresponding increase in crime.

In the United States this has not been confined to any single area. Street crime is the most obvious kind, but then there is also organized crime — probably a much greater threat to government and social order — and political crime, noted in the corruption of law-makers through bribery and other forms of malfeasance.

In the Soviet Union, the criminal element is in charge. It is necessary only to notice the deliberate use of falsehood to influence not only the nation's supposed enemies but also its own citizens. When the central government, which dominates everyone, makes use of the mass media for its own purpose, the deliberate spread of false information is bound to be successful.

The witholding of information is as damaging as the the misrepresentation. Evidently the Russian public, for example, has not been informed that the army invaded Afghanistan and is currently in the process of reducing the people of that country to submission by employing all of the weapons available to a modern state: not only soldiers but also tanks, helicopter gunships and even chemical warfare, all against natives who are still in a primitive technological condition and able to call on nothing to help defend themselves but a few outdated rifles. Nothing in the civilized world as measured by all of the previously accepted standards could justify such cold-blooded butchering of a helpless population.

The great to law and order by so many criminal elements is augmented by a new and dangerous efficiency which is directly attributable to the enormous gains in technology. Weapons of destruction are now much more lethal, and that same technology which gave to those in charge the machinery to control dissident populations could just as easily be used against them. Suppose that some terrorist group, the Puerto Rican F.A.L.N., the Irish Republican Army, or the Palestine Liberation Organization, were to get hold of an atomic or hydrogen bomb. It is hard to imagine how much havoc might result.

In short, there is the grave and very present danger that the same technology which produced all of the benefits of a modern civilization might be equally efficient in bringing that civilization itself to an end. No one doubts what the effects of a nuclear war between the Soviet Union and the United States would be. Each now possesses the capability of destroying the other and in addition of ending all life in the northern hemisphere. Despite the obvious fact that no nation would emerge the winner, such a conflict might still take place. All that would be necessary to produce it would be the conviction that a first strike could be successful. This is unlikely, but the prospect might prove just enough to bring it about.

The Soviet Union at the present time is a nation on a permanent war footing, a country which in every respect is preparing for world conquest. Indeed there is no way that the actions of the Soviet Union around the world can be justified merely in terms of self defense. No country in Africa was a threat. The fomenting of revolution in so many underdeveloped countries can be interpreted in no other way. How else then could the fact that the Soviet Union has been supplying any and all dissidents with arms and ammunition be otherwise interpreted? Never in peace time has a general hostility to all other established governments taken such active form. So far as the Soviet Union is concerned, the world is divided into those countries it dominates and those it opposes, a policy which leaves only the thinnest of lines between peace and war.

5. Belief and Technology

Politics is in a way simply the response to the demand for social order. Properly conducted, a government confines its efforts to enabling its citizens to do as efficiently as possible whatever it is that they want to do. It smooths the way for the manifold of activities on which they engage by establishing law and order and carrying out a consequent set of priorities.

Now, as it happens, what the citizens want to do varies from culture to culture and from civilization to civilization. Moreover, what they want to do is dictated to a large extent by what they believe and by what activities their external circumstances allow to them. We can see how these various factors operate if we compare primitive cultures with civilizations.

All primitive cultures are characterized by two tendencies: an excess of beliefs and a defect of technology. *The Golden Bough* is J.G. Frazer's testimony to the heavy weight of beliefs carried by most tribal cultures, and any field anthropologist will bear witness to the inadequate nature of their artifacts. What they lack in material invention they make up by the activities of the imagination. They believe in a whole host of natural and supernatural phenomena, in many magical persons and events; but with the limited ability at their command to shape materials, they can do very little.

The advance of civilization, which is also a result of an increase in population, reverses these proportions between belief and invention. There does appear to be a constant function here, for now as technology increases there is an accompanying decrease in beliefs. Life becomes governed by artifacts, not myths. Civilized man relies more upon what he can do for himself through his increasing control over natural forces, and less upon what he hopes supernatural forces can be invoked to do for him. The increase in the production and distribution of sophisticated tools delivers him over to his own responsibility; he becomes more active on his own behalf and less dependent upon gods and spirits.

In sum, civilized man is characterized by *doing*, primitive man by *believing*. In taking on the attributes of civilization, he has exchanged a passive for an active role, and so looks forward and outward rather than backward and inward. The future could be a bright one if only he could learn to match his rapid increase in technology and the artifacts it provides with an accompanying political acumen. He has made the scientific advances but he has not yet provided the requisite political adjustments, and this can be a threat to his whole enterprise.

Technology is the more modern version of what passes in the older theories of political economy for 'property' and 'the means of production'. It is where we must learn to look first if we would understand the social relations of government and the state. For example it explains as much about what the Russians believe as it does about what the Americans practice; both betray deep-seated approaches to reality.

There is always a connection between the theory of reality of a society and the direction its government takes, a connection which is mediated by layers of artifacts. The link between the state and its metaphysics is a strong one, and no less so because it is not recognized or acknowledged in most cases. We have seen the connection in primitive cultures where beliefs are prominent and technology tenuous, and in civilized states where the technology is prominent and beliefs tenuous, but what is involved is not an explicit matter of beliefs; indeed it underlies those beliefs and furnishes their ground; as technology it constitutes the reality of politics.

CHAPTER XIV

ART

1. Science and The Fine Arts

When such a new and powerful institution as science arrives upon the social scene, there is nothing in the whole culture immune to its effects. The enterprise we should expect to feel it least, if at all, is art. We would be wrong. For art is no exception: the artist knows, even though he may not wish to admit it, that he cannot hope to compete with the brilliance of the recent discoveries in physics and biology. The result is that he hates and fears science and thinks of himself as working apart from it and even in opposition to it. Yet because he lives in a period in western culture dominated by it, he remains under its influence, and he might be surprised to learn how much it has determined what he does.

Art has no effect on science — how could it have? But science and technology have had immense effects on the arts. And when they do it is not through the discussions of art on the part of some scientists but rather through the artists' choice of subject-matter and the method of treating it, different of course for every art.

The most obvious and familiar example of the effect of science and technology on the arts has come about in the art of the theater. The invention of the motion picture film has extended its range almost without measure, thanks to the wide screen and the television set. These have not only made for improved reception but also brought the dramatic art within the reach of everyone. Millions all over the world are now able to view the same drama in public theaters and private homes. In addition to this extension of the art, there is also the immense value of preservation. Libraries of prints now make a permanent possession of many excellent performances, for if the films show signs of wearing, they can always be copied and in this way kept forever. Music, painting, sculpture, architecture, for example, each has its chosen materials and its own method of treating them. But what all have in common is that all works of art are material objects shaped by deliberate efforts in artistic technology, more specifically, by various embodiments of one valuable kind of false knowledge, and science has given a new turn to this procedure.

Of course when I talk about the influence of science on art, I do not intend to imply any deliberate scientific effort. The scientist is unconcerned with art, and the professional artist concentrated only on his own work. It just happens, however, that the scientist like the artist is an individual, and no individual's social isolation is complete; each must be sensitive to some extent to the values and tendencies of culture in his own time and place. The artist by definition is more sensitive than most; and so it happens that many subtle forces beyond those he recognizes play upon him.

In western civilization in the twentieth century, science is, so to speak, in the air. News about it fills the mass media, and demonstrations of its results are to be noted everywhere. The artist is no more immune to these influences than anyone else. And perhaps the most silent and unseen of all is what the sciences themselves assume: the atmosphere of an empiricism which lays down that what can be shown to sense experience, or to some instrumental surrogate of the senses, has demonstrated its authenticity.

Science has affected art in a number of different ways. We see the influence of empiricism in at least two of these ways: in the movement to return to an unaltered nature, and in the attempt to achieve something akin to scientific abstraction. The first represents the effects of an unqualified empiricism, the second those of its instrumental surrogates, I plan in this chapter to show both of these developments in some of the fine arts, more specifically the changes in aesthetics which have come about as the result of the impact of science and technology on art.

2. Art As A Return to Primitive Nature

One of the broadest expressions of this influence is the movement in the arts of a return to nature. Art ignores the basis fact that the distinction between the human and the natural is false: man is a natural object, and therefore anything that he does is natural. The human domain is simply a segment of nature. What the artist admires, then, is not nature merely but, more accurately, undisturbed nature, a nature untouched, and consequently unaltered, by man. The artist feels it his obligation to remind us that nature without man preceded nature with man and is therefore more fundamental because earlier and more primitive.

The argument is not an altogether valid one yet expresses something of value. For the forces of nature are more powerful than any forces available to man, and this is still true even now that man has discovered nuclear energy. The artist feels that by getting in touch with nature as it was before man he is getting in touch also with those elemental forces which carry with them the immense energies he so much wishes to emulate in works of art.

The strongest early impact of empiricism was contained in the canvases of Cézanne. His version of the analytic method was to break up the image into its component segments by means of a number of small irregular planes.

Numerous quotations from Cézanne can be adduced in support of the contention that he thought in scienctific-empirical terms. He described his method as 'experimenting' (1). His 'meditations, brush in hand' (2) were intended to deal with his 'sensations in the presence of nature' (3). He was the empirical artist, observing in nature the forms of geometry. The artist, he thought,

> must see in nature the cylinder, the sphere, the cone, all put into perspective, so that every side of an object, of a plane, recedes to a central point. The parallel lines at the horizon give the extension, that is a section of nature, or, if you prefer, of the spectacle which the *Pater omnipotens aeterne Deus* spreads before our eyes. The perpendicular lines at the horizon give the depth. Now to us nature appears more in depth than in surface, hence the necessity for the introduction into our vibration of light, represented by reds and yellows, of enough blue tones to make the atmosphere perceptible. (4)

Indeed 'everything in art', he wrote, 'is theory developed and applied in contact with nature'. (5)

That his conception of art was influenced by what he knew about the scientific method can hardly be doubted. He was a very great artist, as his work amply shows, and he pioneered a new direction in painting. He had many followers who are well known, notably Braque and Miró, but especially Picasso and the influential Hans Hoffmann. By adding a kind of self-conscious naiveté to their work later followers produced a much more childish version of the movement, one that climaxed in Dubuffet.

It may not be all that many years but it is a far logical distance from the paintings of Cézanne to the drawings of Paul Klee, though both represent the products of the impact of empiricism on studio art. The art of the great masters was done under no such compulsion; and if the work of Cézanne has survived it, that may be more of a com-

1 Gerstle Mack, *Paul Cézanne* (New York 1936, Knopf), p. 323.
2 The words are those of Emile Bernard from Ibid., p. 306.
3 Gerstle Mack, *op. cit.*, p. vii.
4 Gerstle Mack, *op. cit.*, p. 377-78.
5 Gerstle Mack, *op. cit.*, p. 364.

pliment to his abilities than to his theories. On such a representative artist as Klee the effects of empiricism were quite different.

The drawings of Paul Klee often resemble those made by children, and they look too like some of the drawings of prehistoric man found in the caves of Altamira, Spain. The child and the primitive have much in common. But with Klee the effort was deliberate and self-conscious. He wrote,

> It is a great handicap and a great necessity to have to start with the smallest. I want to be as though newborn, knowing nothing absolutely about Europe; ignoring poets and fashions, to be almost primitive. Then I want to do something very modest; to work out by myself a tiny, formal motif, one that my pencil will be able to hold without any technique. (1)

Why did Klee have this ambition? Why the frantic effort to condense, why the desire to work in what he called his 'shorthand'? (2) Can an educated artist really put aside what he has so carefully learned and really forget all about Europe? Why would anyone want to, unless he sought to repudiate the gains of civilization and return to the primitive and the childish. Klee said that he wished to be primitive, but he rejected the childish element, for he referred later in another place to 'the legend of the childishness of my drawing'. (3) That left the primitive, a resemblance he endorsed.

To be faithful to an undisturbed nature, is what the truly primitive artist must have wanted, for he had no other choice when there was nothing else in his environment in the way of human artifacts to disturb him. I doubt that any of the great artists of the past would have had such an aim. Those responsible for the paintings and sculpture of Greece and the Renaissance certainly did take pride in the elaborate clothing of their subjects and the equally elaborated interiors against which they were posed. No return to unimproved nature there, only the attempt to embellish. The ambition to be primitive is one no member of any primitive culture ever had; indeed it is a sophisticated and even a decadent aim, borne perhaps of a diminishing force and an ebbing talent.

Let us see next if what was true of painting was true also of music.

The return to nature in the art of music is clearly represented by program music. The term itself was first employed by Franz Liszt to describe the program notes which were affixed as a preface to a piece of music to guard the listener against the wrong interpretation and to guide him toward the right one. A more accurate definition refers to the music itself, and defines it as any musical composition 'based upon a scheme of

---

1 Paul Klee, (Museum of Modern Art, New York 1946), p. 8.
2 Paul Klee, *The Late Years*, 1930-1940 (New York 1977, Serge Sabarsky Gallery).
3 *Paul Klee on Modern Art* (London 1948), p. 53.

literary ideas or mental pictures, which it seeks to revoke or recall by means of sound'. (1)

Program music, in a word, seeks to move closer to nature by its musical description (or if you prefer, description in the musical language) of some actual concrete situation which exists in nature. Apart from isolated early examples, such as the famous festival of the Pythian games held in 586 B.C. at which a composition was said to have illustrated the combat between Apollo and the dragon (2), program music did not reach its full development until the eighteenth century in Europe. The recognized starting point for program music is generally acknowledged to have been Beethoven's pastoral symphony, the 6th, particularly its second movement which clearly depicted a summer storm.

It has been well said that, apart from the power and originality of Beethoven's compositions, he set norms from which it has been difficult for other composers to diverge. Most of the great Europeans after his day employed the style. By the middle of the nineteenth century, Mendelssohn, Schumann, Berlioz and Liszt had followed it, and at the end of the century Debussy and Richard Strauss. (3) In the music of Liszt, as in that of his predecessor, Berlioz, 'the program does not tell the story of the music but runs paralled with it – an evocation, in a different medium, of analogous ideas and similar states of feeling'. (4)

There are doubtless still other courses program music can, and will, run. It is at its most vivid perhaps in Richard Strauss's tone poems, for example the 'Sinfonia Domestica', in which he sought to reproduce the household sounds of a family, and in Serge Prokofiev's 'Peter and the Wolf', which has been aptly described as 'a luminous fairy tale', and in which we hear the noises made by the barnyard animals.

The sudden and wide popularity of program music appearing in western Europe at the time when it did is consistent with the notion that it represented in music the influence of empiricism. Music was no longer removed and remote but in fact had become a detailed, vivid and ever-present commentary upon the contemporary scene. Composers were no longer content to reflect in their music upon their own abstract domain and to make contributions to it, as in fact they and the mathematicians had done for so many millennia, but were suddenly bent upon participating in the social and natural environment in which as human individuals they were immersed.

Science was a seventeenth century development, but only a hundred years – not much in the life of a culture – were needed for it to reach the composers and to exercise a

---

1 Percy A. Scholes, *The Oxford Companion to Music* (New York, 1943, Oxford University Press), p. 757.
2 Donald Jay Grout, *A History of Western Music* (New York 1972, W.W. Norton), p. 4.
3 Scholes, *Oxford Companion to Music*, p. 757.
4 Grout, *History of Western Music*, p. 4.

heavy influence on their work. That such cultural phenomena are not explicitly recognized and acknowledged tells us nothing about their power. What permeates everything so silently and unobtrusively may be for that very reason the most potent of influences. Abstract or 'absolute' music, as we will see presently, represented a far older tradition; and by the time science made its appearance must have stood for music itself, without much hint that there might be another kind which also could be taken seriously.

Like the response of composers of music to the rise of empiricism, the modern dance was in its way the dancer's response. In place of the five classic positions of the formally established traditional ballet, the modern dancer saw no need to defy gravity but instead responded to the natural movements and needs of the body and endeavored to stay whithin those structures. Loie Fuller thought of the dance as 'the natural reflex of the body to ideas', and she imitated ordinary phenomena: flowers, flames, insects. (1) No ballet slippers then, no rising on points, but bare feet solidly planted on the ground and movements more in conformity with what physiology suggests. Merce Cunningham thought that the dance movements should repeat the jumble that experience receives from life and not make a formal rearrangement of it. In this he had the support of John Cage, with whom he worked extensively.

The same movements I have shown in painting, sculpture, music and the dance can be observed also in architecture. Architects returned to primitive nature because they were led to do so by the technology of the new materials. At the end of the nineteenth century and the beginning fo the twentieth, a radical change occurred. It was probably in the air at the time, as cultural movements of any sweep usually are; but if it can be said to be due to any one man then William Morris deserves the credit. There were paradoxes at work here. Morris pleaded for a return to handicraft and a sort of supposed medieval simplicity, but what he got instead was the complex machine and its products. Yet he had sought above all honesty and truthfulness in design, and this led men back to a consideration of what was most logical for the new materials with which they were dealing.

Pevsner has noted that Ruskin spoke for the whole of the period and for many previous ones when he declared 'ornamentation is the principal part of architecture' (2), and pointed out that the 'great principle of the Gothic style' was 'to decorate construction'.

But a great change was coming, and it was introduced by the new technology of steam and electric power, which, its advocates declared, 'will have something to say

1 Don Mcdonagh, *The Rise and Fall and Rise of Modern Dance* (New York 1970, Outerbridge and Dienstfrey), p. 16.
2 Nikolaus Pevsner, *Pioneers of Modern Design* (Penguin Books, 1960), p. 19.

concerning the ornamentation of the future' (1). The new architects understood what was happening. The converts were many: Otto Wagner, Adolf Loos, Louis Sullivan, Frank Lloyd Wright and Henri van de Velde.

Auguste Perret (1874-1954) stood out as the greatest French architect of his generation. He understood the aesthetic importance of structure and the demands of building materials.

The American, Russell Sturgis, spoke for them all when he wrote that 'if the architects were to fall back upon their building, their construction, *their handling of materials* as their sole source of architectural effect, a new and valuable style might take form'. (2)

It did. Adolf Loos wrote that it is from the engineers that 'we receive our culture'. (3) Steam power, iron, steel and concrete were to call the turn in all subsequent design. 'Why should artists who build palaces in stone rank any higher than artists who build them in metal?' van de Velde asked. (4)

Stone Age man had been confined in his designs to what could be allowed by the intractibility of his materials. When stone was worked by tools of stone and bone, it could be altered only so far. Architects at the turn of the twentieth century found themselves in much the same predicament; they had unfamiliar materials of iron and concrete, and wisely decided to let these determine what form the new buildings would take. The result was the Bauhaus architecture and design generally fostered by Gropius, which had a far-reaching influence. To the battle cry of respect for materials, the genius of Frank Lloyd Wright added another and important one: form follows function. Determine what you want a building to do, and then design the building to do it. The result was beautiful, as machinery constructed in much the same way was beautiful. The influence of technology had reached architecture and was calling the turn.

3. Art As A Reflection of Scientific Abstraction

The artist's naive conception of science is that it consists in a set of abstractions because it formulates laws which are general and moreover does so in an abstract mathematical language. That this was a mistake never occurred to the artists. The mathematical symbols employed in physics have a material reference which they do not have in pure mathematics. But this would have been too much of a distinction to ask the artists to comprehend; it was an understandable conclusion to assume that only science as such is abstract.

1 Ibid, p. 26.
2 Ibid, p. 29. Italics mine.
3 Ibid, p. 30.
4 Ibid, p. 29.

How does art reflect the influence on the artist of scientific abstractions? What does he have to do in his art to signify such properties? Let me attempt to answer these questions by discussing one art at a time, beginning with painting.

The fractured images of the impressionists and post-impressionists we looked at in the previous section were yet to suffer greater harm. The logical end of that line came with de Kooning, who, according to Brian O'Doherty (1) 'burst on the scene in 1948 at the Egan Gallery with 'those electric paintings of white lines swerving and rebounding in the darkness as if tracing the courses of charged particles'. The analogy to atomic physics was an apt one.

The paintings of de Kooning reached the stage of the fast-disappearing image; all but destroyed, it was broken up as far as was possible. But earlier he had taken another turn: the avoidance of the image altogether. Abstract Expressionism sought the ultimate in abstractions resembling those the expressionists thought to be the work of modern mathematical physics, and presented instead a series of plastic formulas.

There is something about the work of de Kooning which suggests that the scientific abstractions may also be a return to nature. Desmond Morris taught a chimpanzee to paint (2) and there is a remarkable resemblance between the resulting canvases and those of de Kooning. The conclusion can of course be read either way: the chimpanzee paints as well as de Kooning, or de Kooning paints no better than a chimpanzee. To defend him on the ground that the resemblance is a virtue is to accept without qualification the return to nature at its most primitive, and inadvertently to reject all of those values we have learned to associate with an advanced civilization.

There was more progress to be made by other painters toward the elimination of the image. This can be illustrated in the work of two men. In the canvases of Jackson Pollock and those of Mark Tobey it was still possible to suppose that one perceived an image, or rather a bewildering profusion of images, piled one upon another in an almost deliberate disarray that left the emotions altogether and substituted a lower order of decoration.

But it was in the paintings of Rothko that the ultimate achievement of Abstract Expressionism was finally reached. Canvases in shiny and dull black separated by a strip of yellow or in horizontal color were guaranteed not to represent anything. Defenders and expositors of the school of Abstract Expressionism speak of 'tensions', of 'the impossible', and of 'the exploration of space', but the argument is hard to make that a canvas can justify itself as a painting yet represent nothing more. With patches of color chiefly rectangular in shape though kept carefully somewhat irregular, it is difficult to interpret or identify them, or, worse still, assign them any meaning.

1 *Object and Idea: An Art Critic's Journal* (1961-1967) (New York 1967, Simon & Schuster), p. 90.
2 Desmond Morris, *The Biology of Art* (New York 1962, Knopf).

It would I think be fair to argue that the movement in art which under the influence of science undertook a return to nature as the analysis of natural forms finally ended by avoiding those forms altogether and even denying nature itself in favor of abstractions which on everyone's view stand apart from nature in the world of logic and mathematics. By no stretch of the imagination can geometry be limited to nature; it is instead an ideal abstraction approximated in nature, which is a different thing altogether.

Rothko represented, then, the reduction to absurdity of a movement which began with Cézanne. Mathematical physics is the language of science by means of which it describes the conditions of the physical world; but it does not pretend to be the whole of that world. There is a concreteness in art to which artists are unfaithful when they rob it of content in an effort to make it as abstract as science. It is difficult to avoid the suspicion here that there has been the perpetration of a gigantic fraud.

The question could be raised of course whether the Abstract Expressionists were not representational after all. Such was not their aim, but performance does not always match intention. The chances are that, given a painting, it would be possible to find some segment of nature corresponding to it, and a segment moreover for which man was responsible. It would be possible for instance by diligently searching the medical literature to find slides showing organic cells and tissues that were not in form too far away from the paintings of Jackson Pollock. Again, a few decaying walls of buildings look not unlike the canvases of Mark Rothko.

After all, there can be only so many variations of patterns in two dimensions. What results from human effort in art is not the invention of new forms but of old forms in new combinations, and even these may not be so new after all. Man is a natural animal, and what he makes out of the materials in his environmetn belongs to it as much as he does. Therefore 'new combinations' mean only 'combinations not represented before of old elemental forms' which must at best be limited in number.

The same developments that occurred in painting are to be found paralleled in sculpture. Here the typical names chosen might be those of Jacques Lipchitz and Henry Moore.

Lipchitz, like de Kooning, represented the last attempt to hold onto representation while letting go of it. He had done conventional portraits but they are not his best or his most typical contribution. What marks him off from other abstract sculptors is his attempt to depict the interior of forms, to turn them inside out, so to speak. He describes his method as 'semi-automatic'.

'I call them Semi-Automatic because these sculptures originate completely automatically in the Blind. The form which I obtain in this way is first of all examined by me from a technical point of view. Everything I judge too fragile or not suitable for the bronze is taken away. By manipulating my form in such a manner, a lot of images suggest themselves to my attention. Ordinarily, one image is predominant. This one I choose.' (1)

In an effort to be faithful to his unacknowledged surrender to empiricism, much has been made by the modern artist of his faithfulness to the demands of his material. Obviously, every material has its limitations, what it will and will not lend itself to. But that tells us nothing of those limitations by which the artist must be bound, for he is not confined to any one material. He may if he wishes begin with randomly selected pieces of scrap metal, the kind found in old junkyards. *He* is the artist, not the material; and so he is in charge, and if one material will not lend itself to what he wants done, why then he is free to choose another.

The number of possible materials and methods is large; in fact artists are now utilizing many not available before. Thanks to advances in technology, the painter can call upon new acrylic paints and the sculptor upon new power tools. The plastic artist has a wider range then ever, so that what he can or cannot do is up to him, a new freedom he has acquired recently thanks to the assistance of a new technology.

Henry Moore, an early disciple of Epstein, was most certainly influenced by the semi-abstractions Epstein made in his youth. But while Epstein in his maturity went on to a more sophisticated kind of representation, Moore continued with the semi-abstractions, finally in his later years embracing complete abstraction, as he did in the sculpture commissioned for Dallas in 1979 or the piece made to stand before the entrance of the new addition to the National Gallery in Washington in that same year.

There is some question whether many abstract sculptures are not merely geometric exercises performed in an attempt to imitate those weathered stones and rocks which seem to represent endurance in time as illustrations of natural forces, stones shaped in imitation of unshaped stones. The issue is whether an artist should bend materials to his will or surrender his will to them. Thanks to a misunderstood empiricism, the authority of the natural forces of wind- and water-beaten granite is allowed to determine the proper form and take precedence over the sculptor's own preferences. David Smith went even further in abstraction, for sometimes he surrendered altogether to solid geometry.

Cultural movements are at their most powerful when they go unrecognized. Although no artist has thought of it in these terms, science dealt a staggering blow to art from which the artists in their characteristic intuitive fashion endeavored to recover while at the same time saving something. How well they succeeded it will be for subsequent generations to decide. Such an evaluation is impossible to make in any final terms in a generation so close in years to the scene of operations.

Abstract Expressionism has gone its own way, matched unwittingly perhaps against a strong undercurrent of traditional representation. Jacob Epstein in his maturity (1), and Wyeth throughout his lifetime, held their traditional ground with immensely favorable results without disclosing the slightest effects of the scientific revolution. Other artists have revealed various degrees of that influence, but it was never total.

The influence of technology on art is perhaps just beginning. In addition to electronic music, there are also developments in the plastic arts. We are beginning to see computer-generated pictures and the results of the new technique of holography, producing three-dimensional images with light.

The scientific abstractions, which so heavily influenced the plastic arts in the ways we have just seen, were having their effect at the same time on the musical scene. As is customary in cultural movements, each art that showed the same tendency did so in ignorance of the others; and it is only when they are looked at together that we can observe waves of similar effects and see a cultural movement.

The direction of music after Anton von Webern has had all the hallmarks of serious abstraction. A sharp break was made with what had conventionally been regarded as the primary requirements of music from Bach to Richard Strauss and Serge Prokofiev. Led by Webern and after him by Schoenberg and Messiaen, the basic requirements of melody, rhythm and harmony were given up for 'music' that was essentially athematic. Dissonance joined consonance in the work of Karlheinz Stockhausen and Pierre Boulez. But it has been John Cage, the American, who has done the most in the new direction to free music completely and make it available to random methods of composition. He has worked toward a state of complete openness in which any combination of sounds produced by any means is available to the composer. (2)

Just as Cézanne first broke up the familiar image into its various planes, and T.S. Eliot employed familiar quotations in an unusual association in his long poem *The Wasteland*, so Karlheinz Stockhausen in his Opus 1970 had four players each with a tape recorder 'reproduce fragments of music by Beethoven'. Each player 'opens and shuts the loudspeaker control whenever he wishes'. (3) Many options are left to the individual player in his version of such musical 'indeterminacy', as it was called, so that no two performances would sound the same.

A group of twentieth century composers has done much to further the opportunities provided by the new electronic music. New sounds on old instruments, such as those made by Henry Cowell on the prepared piano with wrapped strings, plucked strings, or strings to which various materials, such as screws, rubber, glass, wood or cellophane were attached (4); and new electronic instruments wholly dependent on electric current

---

1 And later also Marino Marini, Giacomo Manzù and Gerhard Marcks.
2 Grout, *op. cit.*, Chapter XX, *passim*.
3 Grout, *op. cit.*, p. 722.
4 Hugh M. Miller, *History of Music* (New York 1972, Barnes and Noble), p. 180.

for their power; these were certainly in the direction of the kind of abstractedness the composers who followed in the wake of Webern preferred.

In the electronic studio a new range of sounds was produced and manipulated by electronic means, all serving the purpose of increasing abstraction. Perhaps the most sophisticated of the developments in electronic music is the one invented by Robert Moog in 1969 and known as the Moog Synthesizer because it is a combination of all the new electronic methods of producing sounds, together with new ways of assorting and mixing them mechanically. One variety, 'tape music', records some of the conventional sounds in nature, such as bird calls, the noise of traffic on a city street, and breaking glass, and then manipulates the tape by changing its speed, reversing its direction, by cutting and splicing, or by combining two or more of these. (1)

In the dance a similar aim to use abstractions was made by translating abstract sound into dance movements. Bach and Schubert and others were drawn on for scores by Maud Allan. Loie Fuller had added abstract forms by employing lighting effects and the skillful manipulation of lightweight cloths. (2) The most influential dancer was of course Martha Graham, who created an abstract style based on body tensions: contractions and releases. She influenced an entire generation. The modern dance did not try to tell a story — that was left to ballet — but undertook to draw designs in movement that represented nothing but its own abstract self.

Art as the reflection of scientific abstraction can be seen as clearly in architecture as in painting, sculpture, music, and the dance. I pointed out in the previous section that the early period in modern architecture was characterized by a return to primitive nature as the only possible response to empiricism. Now in the second period, from 1919 onwards, the emphasis was shifted to abstractions. Wright was part of this movement, too, but so were many others, such as Corbusier in France and P.P. Oud in Holland. Together they created the 'international style' so widely copied in the metropolitan skyscraper. The tall glass box began to appear in all of the principle western cities but especially in New York where space had been at a premium and where there seemed nowhere to go but up. But it was a style imitated also in populated areas where there was no such restriction with regard to space, as in Houston and — even more noticeably — in Dallas. The city of Brasilia, built in the jungle for the Brazilian government by Oscar Niemeyer, is an instance in point. Here form departed from function and became a stereotype.

Wright had a passion for fitting his buildings into their environment so that they blended with the landscape, but this had not been true of the classic architecture. Certainly Greek temples for all of their beauty did not seem to fit into their environ-

---

1 *Op. cit.*, p. 181.
2 Don McDonagh, *The Complete Guide to Modern Dance* (New York 1976, Doubleday), p. 20.

ment until they had been in place so long and admired by so many generations that this virtue was read back into them. And the same can be said for the Mayan and Aztec cities of southern Mexico and Guatemala, which do not seem particularly native to the jungles.

Abstract architecture could justify its sitting when it could justify itself as architecture. If it was abstract, then very well: it was abstract in every way. But this manifesto, practiced rather than declared, led to a sterile kind of monotony; it did not embrace enough. And so it has come to be superseded, rounded out, so to speak, by receiving an end to add to its beginning and middle.

The period of 'contemporary architecture' featuring the 'international style' had a summing up at the Museum of Modern Art of work done from 1960 to 1980, and in other ways also showed definite signs of a departure. (1) The Finnish architect, Alvar Aalto (d. 1976) (2), together with the American, Louis Kahn, helped to lead architecture out of the international style. Many new directions were evident, brought together by men with a common desire to break out of the older and now well established mold, a transitional period in which it was difficult to determine just what definite direction would emerge.

At the present time (3) (1979) there is an anomalous movement representing a pause in aesthetic development, in which every architect seems bent on doing something different and unique, an effort which rarely produces a recognizable mass result. The influence of science is not yet over, but it has been blunted and fragmented, and no longer shows a single abstract direction.

The aesthetic feelings are logically neutral; some are reconcil lable with reason while others are not. In a day of declining civilization, when perhaps only science is equipped to save itself, architecture may survive as an art because it has the virtue of utility: buildings must be constructed to be occupied, usually for some specific purpose. The scientific influence (if we include empiricism and count technology) has led paradoxically to a rigid concern for aesthetic criteria as the only viable guide to the art of architecture.

1 'Mirrors of Our Time: 20 Years of Modern Building', by Ada Louise Huxtable. *The New York Times Magazine* for February 25, 1979, Section 6, p. 22 and following.
2 *Alvar Aalto* (Architectural Monographs 4), Demetri Porphyrios (ed.), Academy Editions, London 1979.
3 'Transformations in Modern Architecture', and exhibition at the Museum of Modern Art, New York, April 1979.

4. The Altered Theory of Aesthetics

My theme in this book has continued to be the impact of science and technology on philosophy; and since aesthetics is a branch of philosophy, we want to know how science has affected aesthetics. To accomplish this properly I have had to take a rather extensive detour through some of the fine arts. The hope is that the detour has given us a better perspective. It is certainly true that the fine arts may be viewed in a wider context, one which has the advantage of displaying some of its more functional properties; but first we have still another detour to make, fortunately this time a brief one: it will be necessary to review quickly something of the background.

Man himself, I said earlier, has been recognized as a sensitivity-reactivity system of organ-specific needs, interacting with the available environment at various energy-levels. It is with the most complex of these that we will be concerned here. Man's needs, I also noted, are those of his particular organs; and artifacts, materials altered through human agency for human uses, were constructed to assist in reducing those needs; while institutions were organized to manage the artifacts; and finally societies and even entire human cultures (or civilizations) were made up of the interactions of such institutions. The artifacts of art, art objects, become interwoven with the aims and practices of other institutions.

Rummaging among material objects in search of those which best lend themselves to need reduction, one generalization about them has been found to which there is no exception: all material things perish. Some last longer than others, a granite rock formation longer than a mosquito; but none last forever. Material things all perish, yet it has been discovered that there are some properties of material things which they share among themselves to a lesser or greater extent and which do not perish. Among these are, strangely enough, certain qualities.

In studying the interactions between man and his environment we are dealing not with one variable but with two. The recently discovered data of prehistory shows that man is not a constant, and of course no one ever supposed the environment was. Man in his present constitution evolved from prehuman types. While climatic and geological changes were constantly taking place, there were serious changes also in the human animal, some so slow that they were unknown to him. Yet they must have represented a disequilibrium and must have given rise to deep feelings of insecurity.

His efforts to shore up his position accounted for the first alterations in the environment as he strove to accommodate it to his needs. It is likely that art arose as a secondary development of early technology. When men made crude tools, such as chipped stone arrowheads or clay pots, they may have noticed the comparison between those adequately done and those better done. The delight in the difference may have exceeded the utility of either. Hence there could have arisen the notion of things made in an excellent way for their own sake; that is to say, for the sake of the excellence rather than for the sake of things, and the products of this notion could have been the first works of art: beauty produced as a matter of superfluous caring and only afterwards recognized as a by-product of craft excellence. The making of a work of art may be described as the employment of technology in the shaping of materials for just what those materials could be made to mean in themselves.

In a certain sense, art has never been free of technology, which was already involved in the earliest and crudest of the arts. Works of art have always been made out of materials of some sort, and the choice of materials for this purpose as well as the skills requisite to work them were certainly technological. Nevertheless technology never has got the credit it deserves for the role it has played in the arts. Stone can be cut more efficiently into desired forms by those who have studied its grain, and some clays are better than others for firing. After all, someone must have constructed the first violin and someone else must have later improved it. The organ and other complex musical instruments did not always exist and must at some time have been designed.

The two qualities which go a long way toward including all others are those which have been called goodness and beauty. Goodness, remember, is the quality of the bond between wholes, the quality of completeness whereby all things in the world reflect the quality of belonging together. Beauty is the quality of the bond between parts, the quality of consistency whereby everything in the whole world shares something of the quality of the world as a whole. (1) Beauty is of course not confined to art, but the formal recognition of beauty is to be found in those artifacts which are designed to be works of art.

It has been noted that art makes a special connection through the sense organs because the strict feedback from sensory experience calls for greater intensification. The more an individual sees the more he wants to see and the more what he sees means to him. By seeing he learns to see more. And so *pari passu* for all of the senses: the work of great artists improves the depth of our sense perceptions.

Most of the uses of the sense organs are for the reduction of other needs, but there are also those uses which are peculiar to the senses themselves: sight for its own sake, which has as its eventual outcome material objects designed to be seen and for no other purpose – polychrome easel painting, for example. Similarly, sculpture exists for the sake of touch as well as sight, music expands the auditory sense, and so on. The purpose of art is to intensify life, not merely to provide entertainment. That is why art cannot ever be an industry and why entertainment can.

So much for the functioning of art. Now we have to account in much the same terms for its production. What orients an artist toward an interest in a particular material object, a woman's face or a landscape, is that it suggests to him how he himself could make an object exhibiting a quality similar to the quality of the relations between its

1   See above, chapter IX, Section 2.

parts — its beauty — though much more concentrated and shorn of all its non-essentials. In this way he brings into existence a symbolic artifact, that is to say, an artifact whose leading edge is the quality of its beauty.

A symbol, I should explain, is a sign whose principal feature is a quality. It is therefore more likely to evoke an emotion than communication a meaning, or at least evoke the emotion first. The emotion involves an external reference rather than a private meaning, namely, the reference which is symbolically expressed by being captured in an artifact as a work of fine art.

Art, then, is that peculiar form of construction intended to produce a degree of self-conditioning, and is practiced by the artist on the assumption that others will respond like himself to what he has made. After a work of art is completed, the artist takes his place with other appreciators and responds to it much as they do once the conditions under which he was inspired to make it have passed.

Works of art, then, are material objects — artifacts — specifically designed for their feedbak properties. The behavior they evoke is a consummatory response to the stimulus of beauty contained in the work of art.

By aesthetic behavior is meant *the sensitivity of confronted objects, in consideration of their perfection.*

Art is an effort to deepen feeling as another way of responding to the very sources of being, a kind of deliberate self-conditioning by means of symbolic artifacts to make possible a greater penetration of the external world, an experience which is always demanding, often enjoyable, occasionally painful, but always intense and enriching.

Because of its symbolic nature, the work of art represents a wider order, and in the appreciation of art the delight of belonging in a world of such objects produces a feeling in the individual which provides him with a kind of symbolic security: that all is well with the world after all and that he is a part of it, a feeling derived from the work of art in virtue of its properties. Plato described the feeling as one of harmony, and Thomas Aquinas called it a kind of radiance; so that what can be recognized in the work of art is the radiance of harmony.

It is clear now that when we discuss such a broad-ranging topic as art or science we are looking at one side of a many-sided organization, which is human culture, the new environment that man has made and in which he must live. Any change in the fine arts must result in a corresponding change in aesthetics. The influence of science on the arts makes of aesthetics a study of one segment of culture. What aesthetics makes of man is subject to his prior decision concerning what he intends to make of it. There are so many factors at work by now, however, that his intentions either get lost or become modified by other elements on which he had not counted but which affect the grand total.

Art aims to seek values beyond culture in order to bring into the cultural purview the trans-finite elements which are only hinted at in its highest reaches. The laws of design rise superior to cultural differences. But then the scientific method does, too, and science influences art in many ways, as we have noted. Art can either be blunted by such contacts, accelerated, or largely deflected, depending upon how it stands up under their impact. Just how it changes to meet them is what we have been endeavoring to assess.

CHAPTER XV

RELIGION

The accounts of the earliest religions show them to have been involved with technology. I will treat of this involvement first, and only later take up the question of the relations between religion and science; for technologists were already on the scene when civilizations first appeared, whereas science as a fully developed undertaking had no effect until the seventeenth century.

1. Religion Aided by Technology

Prehistory, the period, say, before 38,000 B.C., has left few records, but the hominids were also tool-users, and we do have artifacts older than the most primitive human remains. The earliest man was a hunter, and his earliest religious practice, burial. If magic was an aid to the successful pursuit of game we can safely venture that it had a religious dimension; no doubt belief was involved in both activities but it has left no trace. We do know that men wished to survive by hunting and if possible to provide for themselves beyond the grave through burial, and that in this way and from the beginning artifacts were integrated with life-style, and technology with religious practices.

If we look among the practices of early man for the origins of religion, we can probably find them among the grave sites. It must have been when someone died that morality first intruded itself as a fact which could raise a host of questions but not supply the answers.

If the earliest remains are any testimony, then already with early man the burials were formal, and artifacts were essential ingredients in the practice of religion. The bodies had been decorated with red ochre and were protected inside caves by an arrangement of stones. (1) The bones of animals were found with the bodies and may have been those of live animals sacrificed to accompany the deceased. (2) There was a religious cult involving the cave bear. Many graves contained small worked and decorated objects. (3) In some instances the head of the deceased was treated like an artifact, it was carefully cleaned and washed and preserved as a talisman. (4) Broken skulls were employed in funerary cults, and were even used as cups in religious ceremonies. (5) Ritual anthropophagy was practiced to preserve connections with the dead as part of 'an indestructible need to establish relations with the unknown and the uncanny'. (6)

No institutions is immune to engineering influence, not excluding the institution of religion. Buildings as well as shrines, and even such artifacts as holy relics, became the foci of pious concerns. Mummification, a technique practiced by skilled craftsmen in ancient Egypt and based deliberately on the process of pickling fish, was intended to preserve the body for religious purposes should the soul wish to reenter it. (7) This was a contribution made by technology to religion, one later abandoned when religious beliefs changed.

With the arrival of full-scale civilizations, technology continued to promote the cause of religion. The technology of writing, at first on stone (the Ten Commandments) and later on slabs of clay in Sumeria and Ebla, and on strips of papyrus reed in Egypt, was responsible for holy books. Institutions, it has been observed, are culture-wide; and so the larger the civilization the more ambitious the religion. The technology which makes civilization possible by providing men with the tools necessary to transform the environment also provided for the kind of communication needed by 'universal' churches.

Temples devoted to the worship of the gods and heroic statues of them were among the earliest of engineering achievements, even before those of Egypt. 'Archaeological evidence points to the growth of the Sumerian city round the temple; the surplus (in foodstuffs) was brought to propitiate the god; the land became his land; his priests were the leisure class; and the crafts which did honor to the god marked the beginning of civilization – and of technical progress.' (8) The Pyramids were religious construc-

1 J.M. Coles and E.S. Higgs, *The Archaeology of Early Man* (London Faber and Faber), p. 136. *et. passim.*
2 Ibid, p. 219.
3 Ibid, p. 237.
4 F.M. Bergounioux. 'Notes on the Mentality of Primitive Man' in S.L. Washburn (ed.), *Social Life of Early Man* (London 1962, Methuen), p. 114.
5 Alberto C. Blanc, in Washburn, *op. cit.*, p. 124.
6 Ibid, p. 119.
7 R.J. Forbes, in Charles Singer *et. al.*, *A History of Technology* (Oxford 1954, Clarendon Press), 5 vols, vol. I, p. 266f.
8 Derry and Williams, *op. cit.*, I, p. 89.

tions, as were two of the other Seven Wonders of the World enumeration by Antipatros of Sidon in the first century B.D.: the statute of Zeus by Pheidias at Olympia and the temple of Artemis at Ephesos.

Ordinarily we think of technology as the means to fixed ends. But it is happens that, while this is often true, it is true just as often that the means can influence the end and even substitute for it. The size of a religion is determined by the technology available. Men build their largest structures in honor of their deepest beliefs. It is the engineers and the architects who give concrete expression to human aspirations; it is they who say in terms of worked and shaped materials what men believe and long to express in feelings.

That this happens chiefly in religion is attested to by the ancient Egyptian pyramids, which were gigantic structures intended to insure the immortality of the pharaoh and his court. The greatest engineering feats of ancient Egypt were devoted to religion, the obelisks for instance which were monuments to the sun-god, Ré. (1) We have before us the impressive evidence of the ancient Greek Parthenon and the other Greek temples. Certainly it was faith in Christianity which was responsible for the importation into France by the early Crusaders of the techniques involved in using the pointed arch which made possible the 80 cathedrals and 500 churches which were built there during the hundred years from 1170 to 1270. (2)

If it is true that religions have been aided by technology, it is also true that they can be created by it. I can cite at least one extreme example. On a number of Pacific islands in world war II, in the campaign against the Japanese, the Americans arrived, laid down airstrips of steel mesh, and brought in airplanes to make a base. To make friends of the natives, they gave away cans of C-rations, cartons of cigarettes, candy bars and razor blades. None of these had ever been seen before by the natives, who regarded them as manna from heaven. When the war was over and the American troops departed, the natives tried to reinvoke these gifts of the gods by constructing a runway of mud, lighting it with torches, and even building a bamboo airplane at one end to act as a decoy. Then they set down to wait.

This 'cargo cult', as it has been called in what is now a considerable literature (3), is something of a parody of the older religions, and it is one for which technology was clearly responsible.

Every great technological advance has an immense impact on human culture. There is no institution which can withstand the large degree of change which was for instance,

---

1 L. Sprague de Camp, *The Ancient Engineers*, p. 46 *et passim*.
2 W.H.G. Armytage, *A Social History of Engineering* (London 1961, Faber and Faber), p. 43.
3 Cf. e.g. I.C. Jarvie, *The Revolution in Anthropology* (London 1964, Routledge and Kegan Paul), chapter II to IV; the *National Geographic Magazine*, 145, 706-715 (1974); the *Washington Post* for Apr. 9, 1964, in a dispatch from Reuter's with a Port Moresby Papua, New Guinea dateline.

the result of the discovery first of copper and then of iron and steel and lately of the high tensile metals. Civilizations progress in terms of their technology, as much the product of engineers as of artists; and the institution of religion is no exception.

2. Religion Confronted by Science

There is some question whether in a study devoted to the impact of technology and science on philosophy, a treatment of religion was needed at all. Traditionally, technology has served as the handmaid of religion, not as a danger to it. Although modern science has nothing specific to say about religion it has been to some extent regarded as a threat. For the discoveries of science do have a bearing on religious teachings, and we shall have to look at how its effects have been interpreted.

A religion, broadly speaking, consists in a social institution, a church, based on the insights of a founder, and contained in a creed or revelation carried out by a general morality whose observance is dramatized by a ritual, and overseen, and sometimes even enforced, by a clergy. Most of the world religions, though not all, claim supernatural elements; while science is wholly concerned with nature. To this extent (and it is a large one) there is a conflict. The question to be asked here is, whether a religion within the limits of nature as explored by science is possible.

One notable philosopher of religion (1) expressed some irritation at the growing attention given to the physical sciences. 'I am interested', he declared, 'only in three things: God, man, and the relations between them'. Obviously it never occurred to him that the relations between God and man might be mediated by the material world as studied by science. He had also of course begged a great many definitions upon which there was no general agreement.

Religions are begun by institutionalizing the insights of a founder, but to survive for any considerable length of time they must be sustained by theologies. Now theologies usually involve theories of astronomy.

The first great struggle between science and religion was won by religion when the new and militant religion of Roman Catholic Christianity led everyone to neglect Greek science. For a long time no one in the west noticed that science had been preserved by the Arabs in Constantinople centuries after it was made to languish in Rome.

Christianity survived for centuries with the support of the Ptolemaic system, according to which the earth at the center of the universe was the solid, flat and immovable base of all things, with the planets revolving around it. Heaven was above and hell below, so that all Christians prayed facing upward toward God. This system, which represented the approved Roman Catholic belief, had the support of Aristotle and

---

1   Professor Harry A. Wolfson, in conversation.

because it was reinforced by the authority of St. Thomas Aquinas prevailed for several thousand years.

Two hundred years after Aquinas the birth of Copernicus ushered in a new era. With him came the theory that all of the planets, including the earth, revolve around the sun. Aristarchos of Samos in the third century B.C. had been the first to discover that the sun, not the earth, was the center, but full recognition of the heliocentric system had been delayed for 1800 years by the authority of Aristotle, who had rejected it in favor of the Ptolemaic conception.

In the new theory the earth was assigned a lower place among the planets, and man deposed from his proud position near the summit of creation. The Iranian philosopher al-Ghazzâli in the twelfth century saw the danger Aristarchos could be to religion. In the next century 'Europe followed St. Thomas, while Islam followed Ghazzâli... Ghazzâli was right and St. Thomas was wrong: science *does* shake men's faith in God and undermine religion' (1); or at least it is fair to say that new scientific ideas shake men's faith in the older religions. It is important to emphasize here, however, that this need not mean that science as such undermines religion as such, only that fresh religious insights are required if religion is to keep up with scientific discoveries.

Evidently, the challenge to Christianity contained in the new theory was not immediately acknowledged. The then Pope, Clement VII, in 1530 approved of Copernicus' work, but by 1616 the Church recognized that the new system was opposed to Holy Scripture, and, on the authority of Cardinal Bellarmine, Copernicus' book was suspended until corrected. Copernicus' contribution was formally approved in 1822 and the sun acknowledged to be at the center of the planetary system, but it was not long before Galileo was persecuted for his Copernican views.

The theological aspect of religion after that took a downturn when the Protestant movement endorsed faith over belief. The question of which one of the competing astronomical theories was to be accepted fell into the discard. As a consequence the damage done to traditional Christianity by Copernicus was never seriously assessed.

Another great shock to the established religions had been delivered earlier, this time not from astronomy but biology. If Darwin was right about evolution, then the whole conception of an immortal soul was in jeopardy. An animal which had developed as a risen ape could not be expected to have the properties of a fallen angel.

The most serious challenge came not from science but from the new social and economic political movement of Marxism. It left a vacuum. Marxism assumes that because religions have served as conservators of the interests of the ruling class, there is no such thing as a legitimate religious need. This is of course a *non-sequitur*. The economic aspect is not the only authentic one. People will always be concerned about their ultimate fate, and seek in religion some way of saving themselves from what looks very much like oblivion.

1   L.S. de Camp, *op. cit.*, p. 285.

In short, the impact of the new scientific findings in physics, astronomy and biology, have shattered the old religions and left a vacuum which many new cults and some old practices — astrology for example — have tried in vain to kill. Science does not replace religion but can undermine it inadvertently.

As long as people recognize that they must eventually die, they will be concerned with questions of religion. It has been an unequal fight between the discoveries of a new institution of science on the one hand and the outmoded beliefs and dogmas of the old institutions of religion on the other. Yet there is no reason why new religions should not arise which might be just as compatible with recent scientific findings as the old ones were with the science of their day.

A religious shift must surely follow the second revolution in astronomy which took place when the helio-centric system was replaced in its turn by a galacto-centric one, further downgrading the importance not only of the earth but also of the sun, for galaxies were found to be widely distributed in a universe now recognized to be vastly larger than had been thought. How little the view of the earth taken from the moon has thus far affected our thinking! The impact is yet to come from this visible evidence of the pettiness of all human concerns.

Few see in it, however, the components of a theology, chiefly perhaps because there has been no new founder and no fresh insights on which an appropriate institution of religion could be constructed. It won't do for instance to refurbish an old conception by keeping God in human form; for now he would be so large that he would have to be measured in light-years, with clusters of galaxies for his body and the laws of nature as his edicts. Religions take longer to develop than anyone has suspected. No doubt a new one is coming, because death is an aspect of human life which will have to be provided for.

In the next, and last, section of this chapter, I want to suggest what I think the heritage of science has left for the possibilities of religion.

3. Mystical Materialism

There was more than a hint toward the end of the chapter on matter that the spiritual life of individual man need not be denied by empiricism but can be expected, acknowledged, recognized and understood. (1) I said there that it ought to be possible to have a religious enterprise consisting of the effort of man to get in touch with the dominant inner quality of the universe. Since man himself derived from the objective world and reflects it, a richness of diversity of values exists on both sides of the curtain of consciousness. Those values on the subjective side are a cultural selection made by society from the larger supply which belongs to the environment. The urge to

1 Section 7.

identify the long-range self with an object sufficiently permanent to be worthy of such identification is a religious urge. And the size of the object is an index to the intensity of that urge. The larger the universe the greater the longing for identification with it or its cause. The desire for the good of all the parts or for the beauty of the whole world is spiritual.

Everyone carries approximately the same weight of beliefs. The emotional charge is roughly equal, whether distributed among a few propositions or many. If faith is belief without reason, then everyone has that minimal faith which is faith in reason itself. But by and large people find more support in spreading their beliefs thin and in holding to them absolutely: the less one believes in, the more one needs to believe in it. Great tensile strength is required to find sufficient support in the delicate yet firm sub-scription to proximate guesses. Witness the 'passionate sceptics' of the seventeenth century, witness Hume.

Any new religion which does not contradict the findings of science must recognize the paradox that ultimate knowledge seems to be reserved from us but that nevertheless we seem committed to seek it. We can no more give over our efforts to discover the nature of the cosmos and its causes than we can discover them. The compromise seems to lie in a willingness to settle for the time being for conditioned and limited knowledge. We do know what we know; but our knowledge is not final. Thus the true religious enterprise is inquiry into the nature of reality, an inquiry in which the physical scientists for the moment lead, though they make no presumption to exclusivity.

The chief instrument with which to begin the inquiry is hypothetical reasoning. Tentative knowledge has its own form of expression: the language of probability. Statistical arguments do presuppose a background of hypothetical truths, yet these need not be asserted. The penumbra of assumptions is easier to change than the dogma contained in a creed. In religious inquiry we should recognize that we are dealing with considerations involving possibilities and that the results of the inquiry should be treated in terms of the modal categories. As Rescher has it, the 'function of the modal categories is to provide the machinery for an exact logical articulation of the informal idea of the "relative degree of potential commitment" to the endorsement of statements, reflecting their priority-rating with respect to "fundamentality or importance". (1)

The empirical criterion in terms of which the prospect of establishing a new religion for materialism might be explored is one which can be used to measure the older religions in an attempt to evaluate their accomplishments.

When we look to see what has happened, the first thing that strikes us in that the members of one religion have not fared any better than those of any other. They do not contract fewer diseases, they do not have less poverty, they are not less ignorant. If God has a favorite religion we have been given no sign which one it is. We are

1 Nicholas Rescher, *Hypothetical Reasoning* (Amsterdam 1964, North-Holland Publishing Co.), p. 47.

justified in concluding, then, that so far as religion is concerned God is unaffiliated. Man finds religion comforting because faith is comforting and doubt a source of discomfort. But on this ground (even if on this ground alone), any religion will serve as well as any other, for all faiths are *equally* comforting. Thus no established religion seems to have any right to speak for religion as such in any exclusive sense.

When we ask which people are better off and why, we are struck by the fact that the determinant is not religion but science. Technology, science and its offspring, applied science, and the industrialism which results from these, seem capable of providing a better life in this world than any religion left on its own. Prosperity, longevity, health and education, replace poverty, disease and ignorance, in countries which have fostered science. Judged by these conditions if the world is the result of the will of God, then it would seem the He prefers His people to pursue science rather than religion.

The one blot on this picture is the continuance of the phenomenon of war. Science has developed the instruments of war to a dangerous point, but then religions were responsible for as many wars as any other institution, wars as widespread, as ferocious, and as indiscriminate. On this score, then, there is nothing to choose between religion and science; so far as peace is concerned, there is. The by-products of science through applied science and technology could make for a better world, certainly one with more food and better living conditions generally. And if religion claims that it provides immortality, the empirically oriented could reply that the case is yet to be demonstrated. We know no more of what, if anything, happens after death than we did at the outset of such speculations many hundreds of thousands of years ago.

Materialism is not inconsistent with the belief that some material events, such as the shaping of symbols which point toward the limits of matter and the craving of human organisms for identification with large, far-away and enduring objects, to say nothing of the feelings of exaltation which accompany such cravings, indicate the existence of a passionate curiosity about the whole material universe and its possible cause. All that a new materialism should do (as contrasted with what it does) is oppose the claim of some men to extra-material authority for the exercise of the control over other men. If reality be defined as equality of being, then no material part of the whole universe is any more authentic or closer to the whole than any other part. But also every part is an authentic part. Everything and everyone is saved or damned together, in the intimate and unlimited community of formed material.

What is religious feeling? It can be described as exalted emotions implicit in the craving for long-range persistence, arising from identification with far-away objects representing the size and permanence of the universe, and seeming to bring us in this way closer to its cause. Such a craving is disclosed by the inclination toward a belief in immortality despite the absence of any evidence derived from experience.

It is worthy of note that great art can do as much as can be done to transport the individual beyond his petty concerns and into a glimpse of those values which may exist at the level of the whole universe, and worthy of note also that many of the more successful of the older religions have always taken advantage of this fact, though the implications of such support have never been noticed. Thus a materialist version of mysticism is possible on the basis of what we know about matter and have learned to do with it, a mystical materialism able to justify religious feelings.

We would not have to go beyond the material universe, then, in order to account for religious responses. As with Spinoza, the spiritual can be provided for without designating a special category. It can be accommodated by attention to the activities of materials which in some contexts are directed toward the limits of matter. Some recognition of the immense extent of the universe is all that is needed to make us feel its essentially religious nature.

What lies between the cause of the universe and the universe itself are its limits: what are these, where do they stand and what do they mean? The most religious undertakings currently being pursued are those of the physical sciences, the exploration of the components of the universe and their dimensions and constructions. Here is no idle mysticism invoked by means of the deliberate cultivation of abnormal subjective states, through fasting, prolonged prayer, or some other mechanism for promoting feelings of ecstasy. Here instead is a mystical materialism, produced by facts and elaborated by reasons, on the basis of the objective world as observed by the scientists.

Modern astronomy has made the individual increasingly aware of the immensity of the cosmos. The millions of planets orbiting the suns in our Galaxy, and the vast system of millions of galaxies, dwarf petty human existence and threaten to rob it of meaning. How puny seems man and all his efforts, how feeble, how very unimportant!

And yet need that be the correct reading? The evaluation of a given body of evidence and the determination of just what hypothesis that evidence supports is the most delicate and easily distorted of all empirical undertakings. Could one not for example, come logically to the opposite view? Civilization itself, we learn from archeological and geological studies, is hardly more than 10,000 years old; how very short a time that is compared even to life on earth! We and all our works will in all likelihood be swept away into the debris of the next interglacial loess. Yet does the very fact that the emergence of man coincides with the end of the last ice age not mean that we are looking at an event horizon beyond which we cannot see but which might be concealing a wealth of previous civilizations? Has it not all been done before and the traces obliterated?

The record is far from clear. Given the profusion of nature, that for instance millions of sperm cells are produced for every one that fertilizes an egg when any one of them might have, we should be able to conclude that no individuals count more than others. But if there is not a special destiny in the split-second life of the merest meson, then all was lost from the beginning and there is no sense to anything. The significance of human destiny rides on every infinitesimal shred of being as much as it does on sharing in the fate of the infinitely large meta-galaxy.

There is nothing in the human individual that need perish forever. His cells, his organs, even the peculiar arrangement of the organism which gives him his uniqueness, exist as potential when they are not actual. And so it is not inconceivable from biological knowledge that he could occur again.

It is not through participating in the longest survival but having had a place in history at all that holds the greatest hope. For the shortest existence is as much an authentic existence as the longest. We compete not through activity so much as through understanding, and our brief visit here on earth has the import of intensity if not the grandeur of extensity. We cannot ever achieve an effect comparable to a colliding galaxy but we can know about it, and that is perhaps second only to the event itself.

It would not be inconsistent with materialism to suppose that there is no reason at all why there is matter. For it is entirely possible that being is an effect without a cause. Failing that, where we know the effect and not the cause we can reason from effect to cause and assert that the effect must have been the effect of a cause and therefore that there must have been a cause; but this gives us no license to say whether there was a cause or *what* that cause was if it was. There can be no conclusion to the whole from the behavior of the parts.

It does seem as though the more important a topic is, the less we know about it. Science has nothing to say for or against the idea of a creator. It studies the created world and leaves open the question of whether the world was created and if so how; there is no evidence. Any evidence for the creation of the world would have to lie outside that world in an area where science never ventures.

An explanation of the universe may exist and the human mind may at the same time be incapable of comprehending it. Kant was right that we can experience only what our limited powers allow, and these powers confine us to the world disclosed by the senses and confirmed by the reason. All transcendental knowledge is acquired by means of analogy, and all analogy is limited where applicable at all.

But the best we can do is to argue not only from the world to God, then, but if need be also from God to the world. If God *is* the cause of the world, what is that world like? We can study the effects in order to learn something about the cause. If the world is the will of God and whatever happens does so in accordance with His intentions, then the most religious interest we could have is in the world. We can learn nothing about God through mere speculation on the concept, but we could learn about the world, for, as Galileo pointed out long ago, nature is the book of God.

We are justified in asserting, then, on the basis of our powers, that we mean no more by religious undertakings than the exploration of the limits of the world. God stands at the limits, or, if not, then is represented in our inquiries by those limits. We can say no more, though it is much that we have said when we have said this. Peirce was convinced that a passive and undiscriminating attitude of sensible receptivity would result in 'The perception of... manifold diversity of specificalness in general' and that this would amount to 'a direct, though darkling, perception of God'. His name for it was 'musement'. (1)

The intricacy, the complexity, the unity and diversity of the visible world certainly do point to a completeness and a harmony. What is usually meant by a belief in a god or gods is faith in the essential rightness of things, a rightness upon which man feels he can afford to rely when his efforts to probe into the meaning of his brief and painful existence fail.

Super-naturalism is not truly essential to the conception of God. There is a God of naturalism, too. Activity does not disclose His presence, we are left with sense experience and reason. But God is neither a sense object nor an abstract idea. Since there is insufficient evidence either way, the whole topic must remain in the form of a question.

Those who speculate concerning nature do not give up their activity when their observations and thoughts approach the limits of nature. And the limits of nature may be as close as it is possible to approach to the reason for nature. There is no necessity to consider that the only way to see nature as a whole is to look down upon it from above; there is also an inclusive system to be observed by looking as far as possible along the surface.

If we know anything about the universe, it is that all things in it are connected. The limits of nature are to be found in two directions: toward the largest whole, which is the meta-galactic system of cosmology, and toward the smallest parts, which are those studied in particle physics. Is this not after all what we mean by the holy, that there is a unity to the universe, that all material objects consists in the same set of entities, and that all obey the same laws? Yet implicit here also is the idea that there must be a diversity to unify, that without differences there could be no movement toward unity. And so the unity is not a primal unity with unreal Parmenidean or Hegelian parts, but a unity of a real whole composed of real parts whose differences are as real as their similarities. But if all things in the universe are *equally* parts of the universe whatever other unequal properties they may or may not possess, then this equal participation is what is meant by the holy, and that person is most responsive to the holy who is sensitive to this fact.

The greater part of the time spent on human affairs is occupied with a living, raising a family, relaxing with friends. Religion does not need to be maintained as a special enterprise but can be the determining factor in how we conduct the ordinary business

---

1 *Collected Papers of Charles Sanders Peirce*, Charles Hartshorne and Paul Weiss (ed.), (Cambridge, Mass., 1931-35, Harvard University Press), 6.493, 6.458, 6.613.

of life. A religious attitude rightly understood (which is to say, accurately comprehended intellectually and accepted by feeling) is not a question of how clearly one sees his religion but rather of how clearly by means of it one sees everything else.

It is possible to specify something of the proper approach. The flexibility of belief called for by the new conception of the natural world requires balancing attachment with non-attachment, seeking an unaffiliated truth, practicing active non-interference, maintaining half-belief, having reverence for everything, exercising standpointlessness, and reasoning from the Unknown God. It calls for staying on the positive side, making a dogma of fallibilism, preferring safeguards to rules, and holding no beliefs beyond inquiry.

CHAPTER XVI

CONCLUSION: THE FOREGROUND

1. Technology and Society

We have been looking in this book at human affairs as they have been conducted throughout history, and we have found them to be conditioned heavily by developments in technology. Would it not be good sense to try by extrapolating this relationship somewhat to make an informed guess as to the future of man and the direction of this technology? To ask what kind he will develop and what will be its effects upon him, I admit, is to pose very complex questions, and the first impressions of those who have sought answers are largely negative. (1) But perhaps we do not have the necessary factual information or the mathematical knowledge.

Any attempt to predict coming events except in the most general terms is sure to meet with failure. The future is thick with possibilities and the imagination too weak to cope with facts. All we can hope to do here, then, is to consider a few characteristics and rearrange them, as Lewis Carroll did with the contents of a middle class English drawing room and the little girl who lived in it. No one, a generation earlier than ours for example, could have predicted the moon landing or the nuclear fission bomb. If science has proved anything it is that facts as well as laws exceed anything the fancy could devise. Who outside the sciences could ever have anticipated black holes, anti-matter, or the neutrino universe? Most of the great discoveries come as surprises, and bold indeed is he who would risk a guess about the next ones.

1 George Bugliarello and Dean P. Doner (eds.), *The History and Philosophy of Technology* (Urbana, Ill. 1979, University of Illinois Press).

Long range prediction of what lies in store for human culture and with it the technology and science it has lately produced is therefore likely to be worthless. Things never quite work out as forecast, and for a very good reason: there are so many variables it is impossible to take them all into account. We are left, then, with predictions in the short range, but even that kind of anticipation is questionable enough. Still, given what we know (which is pitifully little) we might venture on some guesses about events in the immediate future.

The prospects for technology and science are questions involved with the survival and flourishing of the culture which sponsored them. At the present time these seem to be unlimited. Given an unrestricted development, the earth could become a base for the staging of space craft intended both to explore and to colonize the solar system. Man may indeed have shaken himself free of his earth bondage. Human ambitions have no limits because, as we have noted, human needs have none. The ambition to achieve all that we want to know, to do and to be, must extend outward toward the boundaries of the universe.

Neither the individual nor the family is a valid isolate because neither can exist alone for more than a brief while. They need the support of society under the rule of exogamy, this is the genetic necessity; but there is also an epigenetic necessity as well: they need the tools and institutions which can provide for future needs, and to this end only social cooperation will suffiece. Sociologists will have to include in their considerations the feedback from artifacts, which substantially alters the social situation. Thus society has to be understood in terms of the technology requisite for the production and employment of artifacts.

But all this is a matter of cultural conditioning; and when the culture is abandoned, as it might be some day in favor of another one with a different set of goals, then technology and science as we have come to know them, will reach an abrupt end. After all, civilizations have been terminated in the past, sometimes as the result of destructive conflicts with other civilizations, at other times due to natural disasters. Ice ages probably did more damage than the hordes of Genghis Khan.

Technology is a global affair, and the challenge to social organization is obvious: it too must be world-wide. Will people discover before it is too late that loyalty to a limited social group is the same in its effects as disloyalty to mankind? Politicians are engaged in the struggle to direct technology and science, and to control artifacts. Is the species, man, flexible enough to make an accommodation with a changing artificial environment? Those who have any workable stock of knowledge operate secretly on the assumption that the certainty of ignorance is the only absolute truth. Could a science of society rest on anything less?

The crucial question is whether the civilization which has brought the sciences to this stage of development will hold together long enough for further advances, or will social revolutions cancel all else? There is no great future for a century whose major decisions are made in the streets. The simple and uncomplicated life, so beloved of some, was possible when there were few people of any kind; it will not serve now. The question of whether nuclear energy is necessary despite its risks if we are to provide for the current large populations cannot be decided by those who do not have the requisite information, only a passion to stand across the path of progress.

Not every alteration of the environment is destructive. Some alterations are improvements, otherwise there could be no such thing as civilized man, only well-meant animals living in comparative isolation. The current fashion is typefied by the Alaskan settlers from the 'lower forty-eight' who wish to survive in the wilderness without disturbing it. We could not all go back to the woods. The neo-primitive who does not want to take any more from the non-human natural environment than he puts back into it may have a valid moral intention but certainly not a long life-expectancy such as the inhabitant of the modern industrial city enjoys, and moreover he does not begin a process that ends by producing a Plato, a Shakespeare or a Mozart.

The current rejection of cities in favor of life in the wilderness may do well enough for a few people, but it can never improve the life style of 215 million Americans and of many more Europeans. Nuclear fusion will probably provide the answer in the future, but there are energy needs now that must be satisfied if everyone is to have enough food, warmth and transportation.

The solution lies not in abandoning an advanced technology but in improving it and maintaining its benefits while trying to solve the problems it raises, such as those of waste and pollution, and the dislocations of an energy shortage. There are many civilized activities which no man of good will wants to share; no one wants war and destruction, only peace and construction. But such goals are not achieved by throwing out the whole of civilization, the good with the bad. The most pressing human problem is how to utilize the constructive advances in technology while eliminating those which lead to destruction. Thus far we have not the slightest hint of how this can be done.

2. Science and Society

Will society sustain science long enough to make it possible for science to develop a science of society? This is the key question concerning the future not only of science but of society as well.

Science, which began as an extension of technology occupied mainly with practical matters, has developed into one of the grand routes of inquiry, concerned with ultimates. The question is, not only 'How can we improve the human condition?' but 'What is the universe like?'

Chronologically, art and religion were among the first institutionalized efforts to propose answers to those ultimate questions which most concern thinking man. They go back perhaps to the beginning of the species some 40,000 years ago. Then about 600

B.C. philosophy was invented as another such effort, and in the 14th and 15th centuries of our era the scientific method put in its appearance.

Art, religion, philosophy and science, are the four grand routes of ultimate inquiry which have in common that they arose sporadically and all sought final answers. The speculation is tempting: where there have been four, could there be a fifth and sixth? Science viewed in this light is only the latest of such efforts; and even though it has produced the most dazzling results, like the others it has failed to provide any final answers.

We still grope in the dark. We have not the slightest idea why we are here, what we should do, or whether we can hope. But, oddly enough, even though the enterprises designed to find answers to these questions have failed, they have been usefully employed: they make possible the operation of institutions. In short, they are treated in practice as though they had succeeded in theory. Civilizations are founded and continued on the basis of unsatisfactory ultimate answers.

All four of the grand routes of inquiry have justified their existence in other ways, for each has exerted profound effects on individual lives and on the shaping of societies. The unexpected benefits have justified the efforts even though we do not know any more than we did before about ultimates. On such questions we are still left in doubt. From the answers proposed there is no evidence that we are on the right track. The existentialists have wisely pointed out that we have clearly been given over to our own responsibility whether we care to shoulder it or not in what at least appears to be a meaningless world.

It seems obvious now from recent big science that the greater the effort put on scientific inquiry in terms of expensive instruments in physics and astronomy, the more the human species learns of its environing universe. There is no limit that can be set to the prospects of further results except those set by society's willingness to make the effort.

So much for the future of science. The future of society is somewhat more clouded. A science of society has not been developed. There are no invariantive social laws; there are no hypotheses stretching across all societies; there are no controlled experiments in sociology and no possibility of social laboratories. Unlike astronomy, where the subject matter is too far away to be reached, in social science it is too close at hand to allow for the proper detachment. Still, it haunts thinkers to remember that whatever has an empirical basis should also exhibit statistical trends susceptible to measurement. Societies are built up from units which in turn are built up from similar units in an unbroken succession from atom to man and beyond. Why then would a method that has worked so brilliantly at all the lower levels not work in this one as well?

Both technology and science continue to move in unpredictable ways. Who at the outset of the twentieth century could have expected some of its most important inventions: the atomic bomb, telecommunications, the computer, the production line, jet aircraft, recombinant DNA, plastics, rocketry and television?

The average citizen is called on to live with machines he does not understand, and just as often to live *for* machines instead of *by* them. A television set simply has to be turned on, a motor car driven, a telephone used. They need constant servicing, and he is helpless when this assistance is not available. He is helpless also in the hands of an advancing technology. Expert craftsmen suddenly find their skills obsolete, like the maker of the finest horse-drawn carriage when confronted with the sudden appearance of the Ford coupé.

Keeping human aims central so that technology serves them rather than being served by them requires constant vigil. The effects of inventions always come as surprises. Nobody planned or wanted the 50,000 deaths every year that Americans experience on the roads from motor car accidents. But then many discoveries come as surprises, too. What explores found on Mars was not what they had anticipated. The uniform background radiation was not what anybody had expected to find with the X-ray telescope. Fleming was not looking for penecillin in his contaminated dishes.

Human intelligence has the task of keeping up with the machines for whose invention it was originally responsible. A machine is a man with a single idea; the way to improve intelligence would be to devise machines with multiple ideas. The task is to build more tools outside the body which can outperform the body. The prospect is dazzling and most certainly will disclose in the future the results of a technology we cannot even imagine today.

3. Philosophy and Society

Philosophy has a past; it has no present, but it may have future. Nothing less than a theory of reality, however covert, implicit and assumed, runs sufficiently deep to serve as the support for the organization of a society.

Technology and science condition the understanding of reality. In order to see this it is necessary only to compare the cosmology of a primitive tribe, which has to depend for its knowledge of the world upon the unaided senses, with that of a civilization which cultivates the experimental sciences. What we accept as reality is a function of the technology by means of which we extend our beliefs. At no other period in history have men programmed time on large antennas to receive messages which may have been sent by civilizations on other planets in the Galaxy. No astronomer would do this unless there was strong evidence to believe that such civilizations exist. Belief, in other words, is not restricted to what is here but has been projected by inference to what there may be elsewhere, and it is this level of belief that offers societies the support of metaphysics. This has been true ever since societies first existed, and it is true now.

Societies may be founded on revolutions and conceived in violence, but they can be maintained only on the basis of some kind of order, and a social order is a structure having assumptions about the nature of reality. If there is any metaphysics, then there will always be speculations concerning it, and that in the long run is what constitutes the viable substance of philosophy.

Societies are composed of institutions arranged in a hierarchy, and every institution has a theory running back to metaphysics. It is the metaphysics of the leading institution which furnishes the basis for the order of the society. Where this is known and acknowledged, the procedures are thicker and the subject matter denser. A high culture, in other words, is one which utilizes philosophy. It was covert in ancient Greece, it was overt in the Middle Ages in Europe, and it was present in western European countries and even exists in the United States today though there the fact is denied and its presence unrecognized. Perhaps when societies have developed a science of society, the fostering role of philosophy will be acknowledged and brought again into the foreground.

All societies are confronted with opposition from both within and without. The principle of polarity, of balanced forces temporarily held in equilibrium, operates in all actual situations. In human life it manifests itself in the ambivalence of motivation, the desire of every individual and social group to help and hurt others, which inevitably issues in contrary actions having opposed effects: exploiting as against protecting the environment, extending life expectancy but also armed conflict, promoting pastoral idealism (that life in rural surroundings is better) but also high culture with the fine arts and mathematical sciences and their technological benefits. In practical matters currently, society gains from global communications and chemically improved agriculture, while at the same time suffering the damage done by the contamination of chemical wastes, acid rains, oil spills, carshing jetliners and disintegrating satellites.

Does this massive dilemma mean that mankind has been operating under a system of ideas which is erroneous because largely self-defeating, or does it mean that a deep pessimsim, engendered by the possibility that the principle of polarity is an essential element of existence, is justified? A review of all of the known facts and laws of the sciences, in the hope of learning from them how the next turn of events should be directed, might be the best step, and it is one that philosophy could take.

Speculative philosophy has as its aim the discovery of what system of ideas would best serve as the foundations of culture, not just any culture but the culture which represents improvements in the future which might make it possible to better the conditions under which human lives are conducted. Need-reductions, which include everything from basic visceral hungers to cosmic aspirations; intensifications of experiences — all are counted in enterprises which can only be served properly if they are based securely. Thus the end of the road we have travelled from the crude technology of early man issues in imaginative preferences in philosophy faced toward the future.

4. Society and The Environment

There are some last words to be said about technology and society. They have to do with a wider sweep of considerations involving technology, society and the alteration of the environment.

The geographical environment has been seriously altered for thousands of years by human activity. Wildernesses are transformed into cities and cities are connected in a network of roads and highways which collectively mean that nothing much in non-human nature remains the same. It is not easy any more to get outside the range of such alterations, not even in the few spots where some isolation still exists: the heart of the Arabian desert, large portions of northern Canada and Alaska, the interior of Brazil. The North Pole sees air traffic daily, and the South Pole is now a sparsely inhabited region of scientists from a number of countries making camps. The entire globe, and no longer merely selected portions of it, is the scene for the activities of man.

What is true of the surface of the earth is now equally true both above and below it. Above it the airplanes, and above them the highflying spy planes and earth satellites. The available environment now includes to some extent the moon and the nearby planets. The bottom of the sea has begun to be explored and will no doubt be exploited for minerals and other products. Even the weather is not exempt.

Climate does change, and civilizations change with it. The changes in the climate of the Middle East since Biblical times is well known. Classic Greece saw a different climate from the one modern Greece experiences. What is now the Rajasthan desert was several thousands years ago the site of the very advanced Indus Valley civilization. How much of this has been due to human activity is not known, but what is known is that since that time and more especially in the last few decades, there has been an appreciable global climate. (1) Just now there is some apprehension of what the emission of carbon monoxide from motor cars will do to the upper atmosphere. The earth's ozone layer continues to be depleted by fluorocarbon emissions, with possible repercussions on health.

In highly industrial-scientific cultures, the rapid advance of technology produces equally rapid changes in social behavior, not only in customs and institutions but also in the daily trivia of the behavior of individuals. Not only do the lives of our parents seems strange to us but our own early life does so as well. What is delegated to history approaches the present from the past and encroaches more and more closely upon it. This has the effect of compressing the present.

The future is always where it was: just ahead, but formerly we have been in the habit of leaving the past behind. Now, however, we conduct our lives in the narrow

1 Carl Sagan et al., Anthropogenic Albedo Changes and the Earth's Climate, in *Science*, 206, 1363-1368 (1979).

time-band which consists in a smaller and smaller interval between past and future. Thus an advancing technology has the effect of intensifying the present, and in this way of changing the sense of reality. Thanks to all of the instrumental aids to our efforts in the present, we seem to live on a high peak which falls off rapidly both behind and before.

Acutally, due to medical technology the life-expectancy has doubled in Western Europe and America. It has gone from 35 years to over 70, but the blazing intensity of reality makes it seem much shorter. Not only has the frenzy of life increased, but also its significance.

Unfortunately, the same technology which offers so many advantages also makes it possible for the few who control the political apparatus to manipulate the many. Citizens in the Soviet Union are no more than puppets in the hands of the masters of the Kremlin as they prepare the nation for an attempt at world conquest. It is difficult to see where this will end, which turn it will take, or how its excesses can be corrected.

WRITINGS ON TECHNOLOGY BY THE SAME AUTHOR

'The Importance of Technology', *Nature*, 209, 122-125 (1966).

'Artifactualism', *Philosophy and Phenomenological Research*, XXV, 544-559 (1965).

'Technology as Skills', *Technology and culture*, 318-328 (1966).

Understanding Human Nature: A Popular Guide to The Effects of Technology on Man and His Behavior (New York 1978, Horizon Press).

'The Artificial Environment', in John Lenihan & William W. Fletcher (eds.), *The Built Environment* (Blackie, Glasgow & London 1978), chapter five.

INDEX